# Mr Darwin's incredible shrinking world

# Dedication

*To Christine, my companion,
conscience and strong right arm.*

# Mr Darwin's incredible shrinking world

Science and technology in 1859

Peter Macinnis

H.M.S. *Beagle*, Darwin's home for much of the 1830s.

# Contents

# A NOTE TO THE READER

This is an inspection of how science and technology were in 1859, how they were changing, and how these changes (in knowledge, technique and understanding) changed the world. It is largely a history of an English-speaking world, but I have not hesitated to draw on what was, in 1859, the latest intelligence from foreign powers, as set out in English-language journals and newspapers published mainly in Britain, the USA and Australia, which were then three of the most prosperous nations on the planet.

I began thinking about this book believing the world changed in 1859 because Charles Darwin published a seminal book. With research came a growing awareness that Darwin's book was as much a symptom of change as a cause, just one of many innovations of 1859. I began writing with Darwin somewhat relegated, and by the fifth draft, I was beginning to babble about wedding guest search algorithms.

At a poorly organised wedding reception with seating cards but no seating plan, once a few people find their seats, the remaining guests have an easier task in finding their tables and seats. They don't need to check occupied places, they see spouses and partners already seated, or friends wave them over. The rate of seat filling increases as the patterns become clear, and as blind alleys (filled seats) are apparent. It seems that 1859 was the year that the seat filling crossed a critical threshold; a year of cascades and chain reactions of enabling discoveries in many fields.

I wanted to tell the story of the changes that happened in 1859. That has remained my focus even as I broadened my scope, pursuing hygienists, balloonists, royal mutants, electoral bribery, tourists, telegraphists, solar storms, frozen chamber pots, royalty, rabbits, dyes and other oddities, wherever they led. To make sense of the changes, I needed to delve into the state of art, culture, education, technology, science, medicine, health and life as they existed around 1859. It was a different world back then, a changing world, a shrinking world.

Most of my data are from publications which quoted sterling values; I have mainly used those. There were near enough to five US dollars to the pound sterling in 1859. On that basis, US wages seem to have been a little higher than those in other English-speaking countries, but they were still not outstandingly generous. For approximate buying power, multiply sterling values by 70 to get the current value. Where 'pound' and 'dollar' are not otherwise qualified, I mean the pound sterling and the US dollar as they were in 1859. That marvellous modern journal of science, *Scientific American,* was a marvellous journal of science, technology and ingenuity, called *The Scientific American* in 1859. I have elected to use the modern name throughout.

Units are mainly quoted as I found them in the original. There are 2.2 pounds to a kilogram (about 450 grams to a pound) and there are about 28 grams in an ounce. For everyday purposes, an inch is 2.5 centimetres, a foot is 30 centimetres, a mile is 1.6 kilometres—I have used soft conversions throughout. I have not distinguished between short and long tons on the one hand and tonnes on the other. Temperatures are given, as in my sources, on the Fahrenheit scale, which I have assumed I need not translate.

*Peter Macinnis*

# Introduction

*A person living two thousand years ago would have recognised the main features of human existence right up to the middle almost of the 19th Century.*

Jawaharlal Nehru, Foreword, *Story of the Indian Telegraphs: A Century of Progress*, 1953.

# INTRODUCTION

In the chill early morning hours of New Year's Day 1859, in London, the printers at *The Times* laboured to bring out the first issue of the new year. As the steam presses roared, completed copies were bundled up and loaded on horse-drawn drays that rumbled off through the quiet streets. The journalists, the editors and the leader writers were long abed; now as dawn broke, the printers could take to their beds as well, just as the distributors rose from theirs.

Breakfasting readers in London were advised in the editorial to expect a year like those of the previous decade. Much would happen in 1859, but nothing would really change, they were told. By day's end, much of Britain could read the same advice, as a growing railway system spread copies across the nation. By noon, the main news from that issue of *The Times* could be read in the capitals of Europe, thanks to an expanding network of telegraph lines.

A fortnight later, the same details would be all over the United States, thanks to a flourishing steamship trade across the Atlantic, American telegraphs, steam presses and railways. A generation earlier, that speed was impossible.

Change was all around the Victorians, but they were too close to the events to see what the changes meant, and what they promised—or threatened. Ideas were spreading faster, and the new communication methods brought places and people closer together. Yet just as the world was getting smaller, fossil discoveries were posing sharp questions about the age of the Earth. The world might be shrinking, but events in history were getting further apart as the 6000-year biblical age of the Earth was stretched to fit far longer time sequences and—as different sequences were stitched together, end to end—pushing the start of life, the rocks, and everything, far back into the distant past.

It is only fair to point out here that the rank-and-file English vicar had a good education, and many with training in the sciences had no truck with the '6000-year' Earth, calculated by adding together the ages of the prophets as listed in the Old Testament. The scientific vicars and geological curates could see the evidence for an age much greater than that, but a vocal minority of laymen and a few divines clung tenaciously and loudly to the doctrine of a very young planet, misinterpreting and misquoting the geological evidence when they could, denying the evidence when no other avenue was open to them.

Most people remained silent. The ill-informed zealots, the political prelates and a few scientists who opposed evolution on grounds unrelated to science blustered; a few of those who understood science strove mightily to explain difficult ideas. Most of their fellows on either side took shelter and hoped the storm would pass. A few scientists raised legitimate scientific objections which would later be demolished. Geology could not be denied.

The geologists ended up pushing the age of the Earth out from the 6000 years that was popular in the 1600s to 4.6 billion years, making the planet more than 750,000 times as old as people had once assumed. This was no minor shift. To put it in perspective, the scale of the change was the same as that between what Jesse Owens could jump as a schoolboy (25 feet, 7.5 metres), and the distance between New York and London (3500 miles or 5500 kilometres).

The leader writers for *The Times* liked classical references to Greek mythology but were not taken with Greek philosophy, so Heraclitus was not invoked that New Year's Day. All the same, he had put it well, some 2400 years earlier, when he said you cannot bathe in the same river twice; that if you come back to a river, the waters have passed on, replaced by new water. And, though observers see a similar river with very similar water, it is no longer the same river. We will see more of the same and the result will be the same, said the editorial, failing to recognise that what the world would really see was more change, as it had done for a decade or so.

Change thrived in 1859, a year when pebbles would be cast in streams, making ripples that would spread, combine forces and drive later changes in unexpected ways. It was a year for foundation stones to be laid, for milestones to be passed, and perhaps even for a few millstones to be hung around the world's collective neck, gently, silently, so nobody noticed.

## PEBBLES IN THE STREAM

On 12 February, Abraham Lincoln and Charles Darwin both turned 50. Each was well known in his own circle, but hardly world famous. By the year's end, they had both taken steps that would leave them world famous, even today.

In the first days of 1859, France's Emperor Napoleon III set in train events that would lead to war between France and its Italian allies on the one hand, and Austria on the other. The war would lead to Italy's unification, setting a pattern for the German states to follow. One battle was so vicious and horrifying that it led, five years later, to the formation of the International Red Cross and the first of the Geneva Conventions.

In London, workers' unions combined to withstand a lockout of building workers by their employers. Nobody gained, the *status quo* was preserved, but the trade unions now knew it was possible to stand up to the employers.

## FOUNDATION STONES

Charles Darwin published *On the Origin of Species* in November. To many, this remains the greatest change of 1859, yet Darwin and Alfred Russel Wallace had announced the idea in 1858. Then again, many of the key notions, like an extended age for the Earth, were progressing independently, based on the same observations Darwin was trying to explain. Darwin's book was a product of its times.

On 25 April, work began on the Suez Canal. This would hasten the use of steamships and speed the transport of goods and people between Europe and Asia. Many telegraph lines were laid during the year, and while some cables failed, the world's desire for faster transmission of news, information, requests and instructions grew. New cables soon replaced the failures. In Britain, the world's largest steamship, *Great Eastern* was launched.

John Brown was hanged at the end of the year for taking violent action to support the anti-slavery cause in the USA. By then,

Abraham Lincoln was already campaigning for nomination as the Republican candidate for President. His election to the White House would lead to a war—a war that would end with slavery being abolished in the USA.

## MILESTONES

In 1859, according to the best guesstimates, the world's population passed the one billion mark for the first time. The half-century-old worries of the Reverend Mr Thomas Malthus about population growth were still remembered, and the threat identified by Malthus had set Darwin and Wallace pondering about how evolution happened. Flint tools which were clearly extremely ancient were being unearthed in France.

Louis Pasteur showed by an elegant experiment that life is not spontaneously generated, setting the scene for the germ theory that would develop in the next few years.

In London, John Snow's map of an 1854 cholera outbreak was printed, allowing all to see the clear pattern of spread—'something in the water' caused the disease.

Around 1859, science became professional and the branches of science diverged to the extent that a good scientist might find himself (and it was almost always *himself* back then) unable to understand the work of another good scientist in another field.

On the other hand, the first women were just beginning to edge their way into medicine in a few enlightened pockets in

By 1859, most cities employed lamplighters to light and clean the lights.

Europe and the USA. Most of them got in through loopholes that were then immediately closed. But they were in.

Gas lighting was common in the streets of many cities and towns, and the first electric lighting was being tested. Later in the year, in the US, Edwin Drake would sink the first oil well and it was hailed as a triumph.

## MILLSTONES

In the longer term, Drake's well, and the ones that followed, would be seen as a mixed blessing, because later in the year, an internal combustion engine (running on coal gas) was developed. By an odd coincidence, Svante Arrhenius—who in the 1890s predicted that global warming would be caused by rising carbon dioxide levels—was born during 1859.

In late January, Queen Victoria learned by telegram that her daughter had, a few minutes earlier, given birth to a son in Berlin. She was a grandmother, and her grandson would be Kaiser Wilhelm II, the German ruler in World War I. It would be a hideous war, in part because, around 1859, armaments began improving faster than ever before.

## SEEING CHANGE

There were other changes in 1859 that we will look at later, but could anybody have spotted the particular developments that would change how we live? Sadly, we have never had a way of doing that. Technology has to develop, mature, and be applied in new ways, and only time can tell if it will work or not.

In 1859, everybody knew that the world was set to thrive on the new wonder material, gutta percha. This is a rubber-like flexible solid, developed from the sap of a tree. It had been around for a while, and by 1859, it looked as though there was nothing gutta percha could not do. In just a few more years, rubber would take over, leaving only a few specialist roles for gutta percha. Aluminium finally became cheaper than gold in 1859, and many hoped to make their fortunes from it.

Other changes were more subtle: the interactions of increasing literacy and education, steam presses and tourism driven by railways helped to change the ways people lived and played. But it was a slow and devious process of change by small steps, and to see it, you needed to stand back and look at the larger picture.

The changes over a lifetime are more obvious. Many people who were alive in 1859 had seen Napoleon Bonaparte flourish, and die, and some of those born that year would still be around to cheer when that other European disaster, Hitler, lost a war and died in 1945.

## Change on a larger scale

A lot can change over a century. The Seven Years' War between Britain and France was well under way in 1759. It was the first reasonable attempt at a world war, and Britain was winning, all over the world. By 1859, Britain and France had been allies in the Crimea and the French were digging the Suez Canal that would later come under British control. Italy and Germany did not exist in 1859, except as fervent dreams, but a long war between Britain and France seemed both possible and likely.

By 1959, Egypt controlled the Suez Canal, Europe had started what is now an unparalleled period of peace after two world wars where Britain had saved France from Germany, and Britain was clamouring to join France, Italy and half of Germany in what was then the newly formed 'Common Market'. Britain's entry was being blocked by the French. Any new war in 1959 was likely to be rapid, worldwide, nuclear and disastrous.

Even nine decades sees a lot of change. A 90-year-old in 1859 had first drawn breath before Cugnot's steam carriage ran along a street in France in 1769, and was alive when the internal combustion engine was developed in 1859. A person born in 1859 might live to see the first jet airliner fly on 27 July 1949.

Napoleon Bonaparte was born in 1769 and by 1859 he had risen, fallen and died, hated by most of the civilised world. Alexander von Humboldt, a scientist and explorer, was also born in 1769. Unlike Napoleon, he was fondly remembered by all when he died in Berlin aged almost 90, on 6 May 1859. By late May, news of his death reached New York. *Scientific American* reminded its readers that as a young man, Humboldt had seen the French Revolution emerge in glory and go down in blood and gloom, the rise of Napoleonism, the crushing of Prussia and the fall of the Corsican Conqueror. Now, once more, armed men were to fight under the banners of Caesar and Bonaparte, ' … and who can tell what the end will be?' On a 90-year scale, change was obvious.

If we consider 80-year spells, the British and their Hessian (German) allies were fighting the colonists in America in 1779, then in 1859, Britain, America and the German states stood by while Austria and France fought over the future of Italy. By 1939 France and England were at war against Germany, Austria and Italy, with the US as a supporting friend, soon to be an ally in the war. Japan, isolated in 1779, opened its ports to outside trade in 1859, and in 1939 was already fighting a war of vicious and bloody colonial expansion in Asia, while the European states looked the other way.

If you take 70 years as the span, 1789 saw the Paris mob storming the Bastille while a brand-new United States of America was still finding its feet. Britain was in the doldrums after losing America, its richest resource, and a few wretched colonists in Australia hovered on the edge of starvation. By 1859, France had an emperor (again) and Australia was one of the world's richest nations, thanks to gold. Queen Victoria ruled a vast and prosperous empire and the US was also gold-rich. Seventy years on, in 1929, a depression, driven mainly by events in the US and Britain, ruined the whole world; the first time our world economy was sufficiently intermeshed to make it all fall at once.

Let's look 60 years either side of 1859. In 1799, Britain and France were using primitive semaphore towers to send signals; by 1859, the world was being linked by undersea cables, and by 1919, radio signals had reached all the way from Britain to Australia in a single jump. Sixty years more would see the first personal computers beginning to emerge onto a sluggish market.

## HOW CHANGING FASHION MAKES US HUMAN

Take any gap of two or three generations and we curiously derived chattering simians will always change. We react to progress, to shared new knowledge, to fashion—and it has been so at least as far back as the day when the early Cro-Magnon people reached the European land mass. Even with no other evidence, experts can date Cro-Magnon sites, right across Europe, to within a thousand years, just from the fashions seen in surviving artefacts. For the first time, clever apes were chattering, gossiping, sharing ideas and becoming slaves to fashion, beneficiaries of fashion. Fashion made us.

We take the fashions that swept the Cro-Magnon world as evidence that these people talked, and shared ideas. But when the means of communication and sharing improve, the rate of change skyrockets. The really big change in 1859 was one almost nobody could see at the time: the world was shrinking as travel, transport and communication became easier and faster. A world economy was now a real prospect, and world travel was almost in the reach of ordinary people.

You could take any year and show how the world changed during a single orbit of the Sun, but across a huge range of endeavours, 1859 was *the* year of megachange. In 1859, ideas sprang forth, old thinkers died, future thinkers were born, living thinkers thought, and new discoveries, materials and inventions came pouring forth. Science, technology, trade and understanding all began to feed each other.

By 1859, ideas, processes, materials, machines and the printed word could be shared both across nations and beyond national boundaries, much faster than the walking speed of a crowd of Cro-Magnon hunter-gatherers. The world was converging and sharing common values, methods and concepts in a way that would have been unthinkable when Queen Victoria ascended her throne in 1837. Around the world, even the most isolated humans were coming closer to the rest of the world in a process that is still accelerating today. The changes we saw, 100, 90, 80, 70 or 60 years after were commonly set in motion by changes and innovations firmly rooted in 1859.

Darwin's book had an initial print run of 1250, but by early January 1860, it was reprinting, and a pirate edition was being prepared in America. There was little copyright protection in those days, but soon that would have to change as new ideas became world ideas in a world culture. The shrinking world brought a new awareness, it allowed new combinations of materials from different places and a faster spread of ideas. The world was on the move, and as the distances shrank, the different parts carried each other along. People could learn sooner what was done by others, and yearn for the best of all possible worlds.

> It was the best of times, it was the worst of times, it was the age of wisdom, it was the age of foolishness, it was the epoch of belief, it was the epoch of incredulity, it was the season of Light, it was the season of Darkness, it was the spring of hope, it was the winter of despair, we had everything before us, we had nothing before us, we were all going direct to heaven, we were all going direct the other way ...
>
> Charles Dickens, *A Tale of Two Cities*, 1859.

By the time his book came out, Dickens' best of times and worst of times, the early days of the French Revolution, were two generations past, but his words remain a fair description of 1859. Communication and travel were all bringing people together faster in 1859, but at the end of the 19th century, speedy railroads and steamships would spread bubonic plague efficiently around the Pacific rim. Progress was all around, but it was often of the two steps forward, one step back variety, and even good changes often carried a hidden environmental cost.

For a few, Darwin's explanation of biology and geology, and even the universe was profoundly shocking and offensive, but for scientists it neatly explained the patterns they had found and seen. A few mean-minded grinders of the poor saw in Darwin's words a justification for their treatment of their 'less fit' (in their eyes) fellow humans. To others, subsistence farmers, hunters and gatherers, serfs and slaves, still insulated from Darwin's words, it meant nothing at all, but the world's intellectual isolation was crumbling. Literacy was on the increase, and ideas were spreading faster.

Around the world, some men and women of good will sought to free slaves, while some of their fellows worked just as hard to create new types of slave under different names. Some of the new slaves were the poor of the slave-masters' races and nations, forced into the workhouse or coerced to work under intolerable conditions for miserable pay. Some of the new slaves were people of other races, hired under one-sided contracts and shipped to strange countries, where they would toil for a pittance. The centenary of the birth of Robbie Burns, Scottish poet and fierce democrat, was celebrated in January, but in many places, the brotherhood of man that Burns celebrated was an empty ideal.

The Industrial Revolution killed men, women and children through diseases that flourished in cramped quarters, through overwork and through pollution, but we descendants have better conditions. We may think their suffering was all worth it because ours is a richer world, but perhaps the people who made our world might not agree with our blurred and self-centred view of their life in the 19th century.

The New Year's Day *Times* editorial was right, in a sense, when it told its readers to expect nothing unusual. There would be more of the same, but that same would include the torrent of change that had been happening all around them, entirely unnoticed—and the rate of change would only get greater.

By year's end, Mr Darwin had offered his bright idea, but by year's end, other people, events, discoveries and inventions had changed the world and human society far more. People would live longer and die differently, knowing more of the world than ever before, thanks to events based firmly in 1859. New ideas and new methods would change things, but the raw materials people used were changing even more.

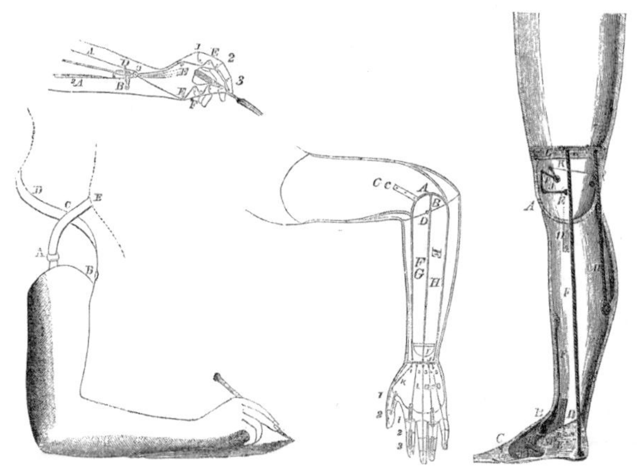

# 1:

# New materials and new ideas

*We had one million bags of the best Sligo rags,*
*we had two million barrels of stones,*

*We had three million sides of old blind horses' hides,*
*we had four million barrels of bones,*

*We had five million hogs and six million dogs,*
*and seven million barrels of porter,*

*We had eight million bales of old nanny goats' tails*
*in the hold of the Irish Rover.*

Traditional Irish song

To modern eyes, this ditty outlines a curious cargo, but in the 19th century, the list would not have appeared so bizarre. If the mills of industry were not as dark and satanic as William Blake claimed, they had a capacious maw, and most of creation was grist for them. Before plastics, before the first oil well and petroleum feedstock, industry relied on the animal, vegetable and mineral kingdoms to fill the world's demands for new things. Nothing was too poor, too mean, too strange to be considered fit for the artisan's bench or the factory's vat.

In a world where literacy was spreading fast, rags were needed for paper-making, lithographers needed stone, horse hides and pig skin might become book-bindings. Bones were turned into glue that might hold books together, dog droppings were used in tanning the hides and dog skins made fine gloves.

Printing was a hot and thirsty job, requiring porter for the workers. It is unlikely the nanny goats' tails were used to tie bundles of books together, but the publishing industry alone could employ all the rest of the cargo, and many other industries could use those items as well.

## SEAHORSE TEETH, GUTTA PERCHA AND DOG DROPPINGS

In January 1859, the screw steamer *Behar* docked at Southampton from Alexandria, bringing passengers, 59 boxes of oranges, 26 cases of seahorse teeth, 728 bales of silk, 490 bales of flax,

23 packages of elephants' teeth, pearls valued at 12,200 rupees and 47 packages of sundries. In late October, the mail packet *Norman* reached England from Table Bay in southern Africa with one package of specie, 253 casks of wine, 2000 horns, six cases of ostrich feathers, 100 cases of arrowroot, 13 bales of wool, 180 bales of skins, 1000 wet hides, 44 tons of copper ore and ten packages of sundries.

Most of these objects make reasonable sense, even today, but seahorse teeth? The OED says the earliest use of 'seahorse' in the sense of the fish scientists call *Hippocampus* happened in 1859: until then, when the term appeared in print, it meant a hippopotamus, or sometimes a walrus—the only way to be sure is to look at the context, as an Arctic origin argues against hippopotamus. Even the famous used hippo teeth: George Washington's false teeth were carved by John Greenwood, not from wood as many people believe, but from hippopotamus ivory, to which Greenwood riveted several human teeth.

Porcelain teeth had been tested in Philadelphia in 1818, but in 1822 an anonymous Boston writer still classed 'the teeth and tusks of the hippopotamus or seahorse' as the best material for false teeth. By 1830, porcelain was making significant progress as a dental material. In early 1860, *Scientific American* noted that porcelain teeth had largely stopped the earlier use of human (or animal) teeth, ivory and bone in dentures.

Elephant ivory found many uses, and at the end of 1859, one pound of the best ivory in New York cost a porcelain tooth maker's daily wage, $1.80. In August, *Scientific American* explained how to soften ivory by placing it in phosphoric acid with a specific gravity of 1.130 (that is, about 20 per cent acid, 80 per cent water). The ivory was soaked until it became transparent, then it was washed and dried on linen. The material remaining was as soft as thick leather, becoming hard in air, but growing soft again when placed in warm water. It could be used to make both nipples for nursing bottles and covers for sore breasts.

Then there was gutta percha, which came from the sap of an Asian tree that could be tapped, much like a rubber tree. In 1832, Dr William Montgomerie saw it being used to make handles for parangs (machete-like knives) in Singapore. He introduced gutta percha to Europe in 1843, where it was used first for knife handles and in golf balls. In 1845, Werner Siemens suggested using it to insulate telegraph wires, the method was patented in 1847, and the first recorded use of the new insulated wires was in 1849.

In 1848, Michael Faraday was delighted by gutta percha. It could be softened by warming, and moulded, but when it cooled, it was flexible and resilient, and it was an insulator which could be employed to seal wires used to carry large electric currents.

By 1853 in the *Gardener's Chronicle*, 'CRD' (alias Charles Robert Darwin, a keen reader) was seeking advice on any problems he might expect while using a canvas hose, coated and lined with gutta percha, as a siphon tube to move water from one tank to another on a different level.

After 1855, the Second US Cavalry's soldiers were issued with a gutta percha talma, a long cape or cloak, extending to the knees, with large loose sleeves. Soldiers in one squadron had gutta percha scabbards and another squadron had gutta percha cartridge boxes, but the material could also serve more peaceful uses.

In April, the *Portadown Weekly News* in Ireland carried an advertisement for 'Ladies' and Children's Leather and Gutta Percha Boots and Shoes'.

Across the Atlantic, an enterprising American was making waterproof packing paper by giving plain paper a thin coating of gutta percha dissolved in turpentine. Two New Yorkers, Johns and Crosby, offered a cement for roofing based on gutta percha: the advertised cost was '5 cents a foot'. It was guaranteed for five years.

By 1853, with the use of steamers, rubber traders were already moving up the Amazon River, so the era of gutta percha would be brief. Critics sniped, suggesting it cracked; that the failure of the covering had destroyed the second Atlantic cable.

Manufacturers of gutta percha cables disagreed, but they still lost out in the end to rubber. And just in time, because the jungles of the East Indies were almost cleared of gutta percha trees by 1859.

A few niche markets were left to gutta percha. A canny surgeon on the Australian goldfields in the 1850s was well aware that gutta percha could be softened more than once, after first use. He told a fellow surgeon he had paid five shillings for a bunch of old tools being sold with a gutta percha bucket, just to acquire the bucket.

With this and a supply of hot water, he had 'stopped' (filled the cavities in) hundreds of teeth at a guinea a piece, and he expected to stop thousands more before the old bucket was used up. He claimed he was known as an unrivalled dentist, with people coming from far and near. Even today, when you have root canal therapy, your dentist may insert a gutta percha filling.

Few trades seem as unpleasant to our modern sensibilities as the work of the collectors of 'pure', the London poor who made a living from collecting dried dog droppings. Pure was used to process morocco and kid leather (putting a new slant on being handled with kid gloves!). Many goods that are now plastic were, then, made of leather which required tanning and, therefore, needed products such as tanbark, which might only be available after being transported long distances. Other materials were more local and in London, the tanners of Bermondsey bought their pure from local collectors, but the practice was known elsewhere.

Times were less fastidious, and *Scientific American* could tell its readers without a hint of a joke that 'fine Russia leather' was well known for its quality and peculiar smell. The hides were first put in weakened alkaline ley to loosen the hair, then scraped on a beam. Calf hides were reduced by dog excrement, a sour oatmeal mash, and tanned 'with great care and frequent handling' using willow and sometimes birch bark.

The skins might then be dyed red with alum and brazil wood or black with acetate of iron and logwood before being dressed with birch bark oil. This oil gave the characteristic smell and kept insects away from books bound with it. Finally, the leather was treated with a heavy steel cylinder wound with wire, to give a barred appearance.

But the days of leather were limited. At Stepney Green in London, artificial leather was being made in pieces 4.5 feet broad and 50 feet long. Mainly India rubber, with some secret ingredients, it was used in book-binding and saddle making.

## ISINGLASS, DYES AND WORLD TRADE

In *Oklahoma!*, Curly sings of a dashboard made of 'genuwine leather'. By the early 1900s, the period of the musical, many imitation leathers were available, but they were just taking off in 1859. Curly's isinglass curtains *might* have been mica, but they rolled down, so we know he meant sheets of gelatin, derived from sturgeon swim bladders. 'Isinglass' is a corruption of an old Dutch word meaning 'sturgeon bladder', but because it was translucent, it was often used as an alternative for glass, among other things.

Isinglass was also used to clarify beer and wine, in sugar refining, as a stiffener in jellies and in some forms of glue. Demand outstripped supply, and William Murdock, the man who introduced gas lighting into factories in the early 1800s also invented 'British isinglass', using the swim bladders of a variety of other fish, including cod and hake. Lantern windows could be isinglass, mica or gelatin made from old bones.

According to a snippet in *Scientific American*, a Melbourne chemist had developed a dye from 'those nocturnal tormentors which infest bedding in many localities, having but one use, "to teach mankind humility"'. There is no indication of whether it was fleas or bedbugs that were the main ingredient, but clearly, no substance was ever left unconsidered, either as a material or to improve a material, or both.

The dyes used on the fine Russia leather were typical of the time. A contemporary recipe for making alizarin ink involved simmering powdered nutgalls in wood vinegar for several days, filtering, adding more wood vinegar, oil of vitriol and gum arabic, agitating over several days, then adding indigo. Nutgalls were growths cut from a 'dyer's oak', wood vinegar was made by distilling wood in the absence of air, oil of vitriol was sulfuric acid, gum arabic was the sap of an *Acacia* tree, and indigo was a dye made from an Indian plant.

Other colours in 1859 came from cochineal (derived from scale insects), and heavy metal salts (mainly lead, chromium, arsenic and mercury). A Parisian perfumer sold makeup based on heavy metals to stage performers of his city in 1859, poisoning a number of them. People were first warned in 1859 that wallpaper coloured green with arsenical salts was dangerous when one wallpaper maker, William Woolams & Co. advertised their papers as arsenic-free. Small wonder that people liked the vegetable dyes and Perkin's mauve, derived from coal tar!

Queen Victoria wore mauve when her daughter, the Princess Royal, married Prince Frederick William of Prussia in January 1858, but mauve really only became widely available in 1859. Later in the year, another aniline dye, originally called fuchsine or roseine, was named magenta to honour an 1859 French–Italian victory against Austria, close to the town of Magenta, west of Milan.

The world and its factories were growing hungry as new distant markets came within reach and new needs emerged. Cotton hauled from the southern US might be woven into webs of 'duck' in New England, which were then coated in rubber from Brazil to make belts that could be sent back to the southern cotton planter for his cotton gins and sawmills. Copper ore from Chile might be imported to Wales to make wire that was covered with gutta percha from Malaya before it was sent to India to be used to lay a cable across the sea to Ceylon.

*Scientific American* put the trade in chemicals like this:

> The never-separable trio, oil of vitriol, soda and chloride of lime, are sent from here [Messrs Tennant in Glasgow] in incredible quantities to all parts of the world. The quantity of oil of vitriol manufactured weekly reaches 600 tuns, that of soda ash 250, of sal soda 180, requiring about 450 tuns of salt a week.

There was also a growing need for fertilisers. Farmers were advised that superphosphate of lime, 'produced by the action of oil of vitriol on burnt bones', was recommended in small doses when transplanting trees. England was spending more than $3 million on manure of various sorts—and various is the key word. A farmer in Little Falls, New York, reported that hair from a local tannery had delivered excellent crops for the next ten years.

Alexander von Humboldt praised guano, accumulated bird droppings from oceanic islands, when he took samples back to Europe from Peru in 1802, but interest only developed towards the end of his life. The US Congress passed the *Guano Islands Act* in 1856 to allow suitable islands to be 'acquired'. By 1859, some 48 islands had been claimed, and extraction was under way. Midway Atoll was discovered in 1859, and annexed by the US in 1867.

The enthusiasm for guano coupled with the relative difficulty in getting it, opened the way for fraud. An article in *Scientific American* told farmers how to spot counterfeit and adulterated material. In order to confirm that they had bought Peruvian guano, they should burn it. The real product would lose 55–60 per cent of its weight, and produce a white ash, which would dissolve readily without effervescence in dilute muriatic acid, leaving an insoluble residue of around two per cent. A bushel of guano weighed about 70 pounds, but if clay, marl or sand had been used to extend it, it would weigh more.

## NEW METALS, NEW ALLOYS

In 1854, Henri Étienne Sainte-Claire Deville came up with a commercial aluminium-smelting process. This drove the French price of aluminium down from around $1200 per kilogram in 1852 to around $40 per kilogram in 1859, still ten times the price of ivory. Napoleon III helped Deville establish a factory, but few people understood what benefits might derive from using aluminium; nor did they care. There was, complained Deville, nothing '... more difficult than to introduce into the pattern of men's lives and to get them to accept, a new material, however useful it may be'.

Aluminium might now be cheaper than gold, it might form alloys, and the metal could be gilded, but its still relatively high price meant aluminium was used mainly for jewellery. Still, the optimists declared the price would eventually fall further, when we would see tea utensils, door knobs, spoons, knives and forks made of the wonder metal. Considering the cutlery suggestions, the writer clearly knew little of the soft, easily bent metal and its properties!

While experimenters found ways to solder aluminium, and techniques to make aluminium pistol barrels and aluminium alloy axle boxes, steel was still the more popular metal. Both the Bessemer converter process and the Siemens open hearth process were introduced in 1856, and by early 1859, the Bessemer Steel Works at Sheffield was turning out high-class steel for tools, cutlery, guns and rails.

Everybody just knew that if enough new materials were explored, some good would come of it. Prince Albert, Queen Victoria's science-mad husband, wanted to promote the use of waterglass, a form of sodium silicate, so he ordered the translation and printing of a German pamphlet on the topic. The pamphlet included methods for preparing four different soluble silicates, and some ideas on how they might be used. This was then picked up by the London periodicals, and so found its way into the pages of *Scientific American*. With no copyright in 1859, the information probably appeared in many newspapers thereafter. What writers lost in the way of intellectual property, society gained in the way of ideas.

## THE PAPER FAMINE

Rising literacy levels caused a huge demand for paper, which was, at that time, made from old rags. Now, there were just not enough rags to make all the paper needed, and the demand went beyond Britain and the USA. Frank Fowler wrote, in 1859, that Sydney's Mechanics' School of Arts '... sends to England some hundreds of pounds yearly for the purchase of newspapers, magazines, and the newest general literature'. The demand from the English-speaking diaspora and a growing British Empire significantly increased the strain.

The use of wood pulp to make paper began slowly in about 1845, and early batches were poor. The price of cotton rags stayed high through the 1850s, so inventors looked around, and assorted chemicals were thrown at vats of straw, seeking to bleach and reduce the stalks to individual colourless fibres. The straw paper was better than wood-based paper and the Philadelphia *Ledger* and New York *Sun* were soon using it. Then sorghum or Chinese sugarcane was tried as a raw material: this paper was reported to be whiter, but it still needed 25 per cent of rags.

By 1859, American inventors had created several machines that made paper bags, so usage there was increasing. But until 1861 in Britain, there was a tax on paper, which led to some curious reactions. One such was recorded when a popular science essayist, Dr Andrew Wynter, described a visit to Price's candle factory in Battersea. The manufacturer used tax-free cheap wood ('deal') for containers, rather than paper or cardboard. A deal plank, one foot wide and four feet long, was shaved into 140 sheets of wood, then thin rice paper was glued on one side of the wood sheets to make hinges after the wood was cut, then the boxes were made up by hand. The sheets could also be rolled into cylinders to provide night-light cases, said Wynter.

Charles Goodyear's battle to maintain his patent for vulcanised (strengthened and more robust) rubber dragged on through 1859. Rubber was beginning to become more popular, and not only in roles filled earlier by gutta percha. Most new uses (hoses, drive belts, seals and shock absorbers—tyres only came later) needed vulcanised rubber.

In 1855, Augustus Gregory experimented with an inflatable boat made of canvas covered with India rubber to explore rivers in northern Australia. It was not a success, as the rubber perished.

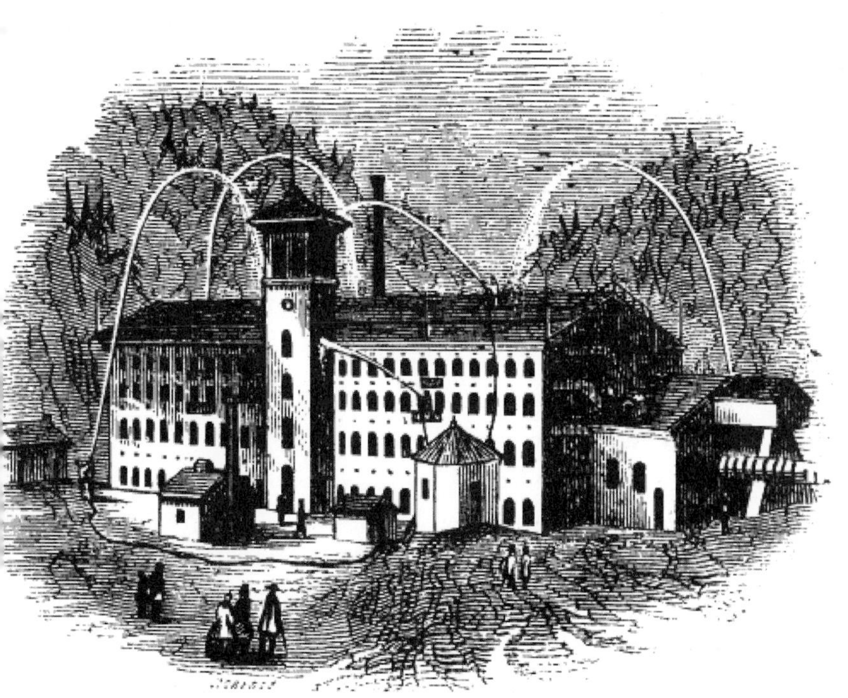

Pressure-testing rubberised canvas fire hoses with steam engines.

In 1859, a diver crossed the Schuylkill River near Philadelphia, striding along in an India rubber suit that covered him up to his neck, with a sheet-copper helmet over his head. He breathed through a pipe connected to a compressor on a boat and followed a guide rope during his 25-minute stroll. On another front, Messrs Badger and Co reported successfully making a Boehm flute from Goodyear's vulcanised rubber, though this was probably more like ebonite than the rubber we know today. Musicians do not seem to have rushed to adopt the flexible flutes.

## Changing textiles

The late 1850s was a time when fire brigades were being formed in many places and equipping themselves with steam pumps. Hoses were a major concern, because the high pressures needed to reach the higher floors of new buildings tore most cloth, and the available range of hoses was poor. Rubber-cotton hoses alone could stand the pressures being generated by the new steam fire engines. The choice of textiles for any purpose was limited: cotton and wool for everyday use, silk and linen for the rich, and that was about it. Wool sold in New York for 11 to 50 cents a pound, depending on quality. Cotton was plentiful and cost eight-and-a-half to 12½ cents a pound. All the same, men mostly wore wool, possibly at a cost to their health.

During an 1859 lecture on hygiene at the Buffalo Medical College, Professor Frank Hamilton warned American men against the broadcloth they had adopted as a sort of national costume. He told them it was a feeble and expensive cloth, giving no protection against the cold of the North, while in the South, it was too thin for winter, too black to be cool in summer and likely to make the wearer avoid activity because it might tear if one mounted a horse or leapt a fence.

Sufferers from bacterial infections applied the spinal marrow of the ox on a piece of cotton rag, changing it every four hours and this, reported *Scientific American*, successfully cured a felon on a Boston lady's finger. This particular felon was a bacterial abscess, not a criminal, though malefactors were also associated with textiles.

Some of them were mixing cotton into more expensive wool, with fraudulent intent and *Scientific American* advised its readers that cotton in flannel could be detected by applying potash (wood ash, mainly potassium hydroxide or carbonate), which converts the wool to soap, leaving the cotton unaffected. Then again, muriate of manganese could be used to dissolve out the cotton in wool–cotton blends known as voile de laine or muslin de laine. This left the valuable wool fibres to be reused.

Another investigator took a different direction. The *London Builder* reported that copper salts dissolved in ammonia made a solvent able to dissolve 'lignine', wool and silk, forming a fine cement and waterproofing compound. This might show the way to more effective insulation of telegraph wires, a reporter suggested.

Fabric could be waterproofed with tallow soap, glue and alum. This was suitable for loose tunics worn over the clothes in rainy weather, but not for clothing itself, said *Scientific American*.

While the Second US Cavalry used gutta percha garments to stay dry, the French army had another way to keep out damp in the Crimea:

> Alum and sugar of lead were combined to give the uniform material a coating of sulphate of lead which repels rain while allowing sweat to pass out, something, we are told, which neither gutta-percha nor India rubber cloth will do.

Steamships were getting bigger, offering larger holds and a more reliable service—and the demand for manufactured goods was increasing. The needs of trade were so great that no nation could afford to lose ships or cargoes in time of war, so privateering, a sort of licensed free-enterprise wartime piracy, where enemy trading vessels were seen as fair game, was banned by the Declaration of Paris in 1856. Signed by all the major European powers, this had the effect of boosting trade in the late 1850s.

## THE GREAT COLONIAL LAND GRAB

The number of rich was increasing fast. There was more capital, there were vast riches to be won in colonies (at a cost to those colonised), and the imbalances between rich and poor were greater than ever. In London, Henry Mayhew was completing his studies of London labour and the London poor. The eminent Victorians led very much easier lives than the urchins who gathered pure or cigar ends on the streets of London.

The demands and prizes of trade were enough to open even the closed doors of Japan. Admiral Perry visited Japan with four US warships in 1853 and pressed for the Japanese to let

US vessels enter their ports. The result was a bonanza, once the Japanese–American Commercial Treaty was concluded. Japan had been isolated for 218 years, with foreigners only allowed to land at Nagasaki.

Now the 'Harris Treaty' made five Japanese ports accessible to US ships. Other treaties with other nations followed, but the Japanese were not a pushover market where cheap goods could be sold at high prices. A British squadron found the Japanese welcomed new knowledge, but had already proved their technological skill. They made good microscopes and telescopes; they understood the telegraph; they made thermometers and barometers including, according to the *Gentleman's Magazine,* aneroid barometers. The sophisticated Japanese were unlikely to rush to purchase shoddy and brummagem, so it was better for British entrepreneurs to create, establish or capture colonies where the people were less discerning.

Colonies provided raw materials and absorbed low grade and overpriced manufactured goods. This was an old British habit: Arthur Phillip, the first governor of the Botany Bay colony complained that the tools sent to Australia with the First Fleet from Britain in 1787 were good only for the Africa trade.

If the empire yielded a stream of wealth, resistance was beginning to firm. The Bengalis were up in arms over being forced to grow neel (indigo) and there were stirrings in Ireland, but the colonials in Australia could be placated with a charade of self-government.

Flexing their puny muscles, the New South Wales Legislative Assembly resolved that Fiji should become a British possession, managed from Sydney. Now the colonies could yearn for colonies of their own, and be content with their lot.

*The Times* reported mid-year, that the King of the Cannibal Islands (Fiji) could not pay a debt of £9000, owed to the US Consul. The king had offered to transfer sovereignty of the several hundred islands to Britain, but the editorialist in *The Times* opposed it. True, cotton could be grown there, but that was so anywhere within 30° to 35° of the equator, and Fiji would make a fine coaling station, but who needed so many islands for one coaling station between Australia and Panama?

The argument that if it were not taken—by Britain—then France or America would snap it up meant little, said *The Times*, or else 'Lord Palmerston should at once take possession of every island and country in the world which is not in the occupation of a civilised Government'.

The other European powers, as equally civilised governments, felt an equal moral compulsion to occupy. Until Germany and Italy became nations, they had few chances of acquiring colonies and, over the next century, this would lead to tensions. France was already snapping up new colonies to make up for their losses in the Seven Years' War and the Napoleonic Wars. In 1859, they were expanding in Algeria while in other parts of Africa, they bought the port of Obock in Somalia from the Danakil clan, to establish a foothold there, and they also took over Rivières du Sud in Guinea.

In Asia, the French force at Tourane in Indochina was weakened by dysentery, cholera and malaria, but Admiral de Genouilly sailed south to attack Saigon. French soldiers entered the city in February

1859, but it took until 1883 to control the whole of Vietnam (which should have made people think a little more closely in the period from 1950 to 1970). Russia conquered the northern Caucasus after they captured the Chechen Imam Shamid in August, and so could force the Chechens out, which they did in 1877–78 (and we know where *that* led!). Timor was divided in 1859 between the Dutch and the Portuguese, another long-running sore that would also plague the 21st century.

In the Ionian Islands between Italy and Greece, a future British Prime Minister, Mr Gladstone, entered into negotiations that would end in the islands being handed over to Greece, much to the disgust of some British admirals who saw the islands as excellent naval bases. The islands had only ever become British, thanks to a comic opera scenario when Napoleon Bonaparte captured Venice, which claimed the islands at that time.

This made the islands nominally French-controlled, so the Royal Navy grabbed them, and Britain continued to hang grimly onto them, post-Napoleon. In Britain, Mr Gladstone's convoluted 1859 mission was itself denounced in *The Times* as 'comic opera', but his recommendations took effect in 1864.

One result of the changes was that Sir George Bowen, the Colonial Chief Secretary in the Ionian Islands, needed a new post. He was sent, in 1859, to govern the newly raised Australian colony of Queensland, accompanied by his wife, Contessa Diamantina Roma, a native of Kefalonia.

In time, they gave their names to the Bowen Basin and the Diamantina Basin, possibly the only two married geological features in the world—if you look at it the right way.

The British liked to believe that they were doing good works, and as individuals, many of them probably were. By 1859, *Scientific American* could report, no doubt on the word of an English account, that India's roads, aqueducts and railways were progressing, and light draft steamers were 'advancing on the waters of her most sacred rivers'. There were bridges as well. 'This is a new aspect in the history of war, to conquer first, then plant—not your standard—but the steam engine, and leave it to work out that truer victory which is gained by the supremacy of the arts of peace.'

Canada was fast approaching Dominion status and in May, the Canadian pound was replaced by the dollar and conveniently, the new Canadian cent coin was an inch across and weighed one hundredth of a pound. In November, the New Brunswick and Nova Scotia Land Company met and recommended that land should not be sold for less than five shillings an acre, with instalments to extend over no more than six years.

Nobody wanted riffraff to come pouring in. Poor and huddled masses were fine, but only if they arrived with sufficient cash reserves.

## MR MALTHUS' NIGHTMARE

The trigger that made Charles Darwin and Alfred Russel Wallace think about evolution was an essay by Thomas Malthus arguing that sooner or later, human population would outstrip the food supply. Malthus made his prediction when rural Britons were fleeing to the industrial towns, but his anticipated catastrophe was delayed because new methods and imports were maintaining the food supply. Still, 1859 saw a billion humans on the planet, the highest number ever—and the rate of increase was climbing.

Just before 1850, the United States' population overtook Great Britain's. By 1860, Britain had about 23 million people against 32 million in the USA, thanks in large part to migration from Britain. Other races were migrating as well: indentured labourers were on the move from India and China and Bantu people were moving into South Africa. The best estimates gave a total African population of 57 to 60 million, much of it in the north, above the 30th parallel.

This was almost certainly an underestimate, but whatever the true total, it would have been an increase on earlier years. According to *Scientific American,* Egypt's population was 2.5 million in 1798, 3.7 million in 1817, 4.25 million in 1847, and 5.125 million in 1859. Alexandria had 30,000 inhabitants in 1798, 230,000 in 1817, and close to 400,000 in 1859. Further south, with the Atlantic slave trade slashed, numbers would also have been rising.

Most growth was in the cities. London was 2.8 million (up from 958,000 in 1801) and New York with suburbs was 1.1 million, while Manhattan itself held about 600,000. London was the place to be in 1859, but other cities were keen to take over, by fair means or foul, subterfuge or even demolition. In Paris, the 'Wall of the Tax Farmers', had been built to enforce taxes on goods entering central Paris. Emperor Napoleon III had it torn down so the areas between it and the outer wall became part of Paris, raising its population by adding eight new arondissements to the capital. Paris, including the suburbs, now stood at 1.6 million, still less than London, but impressive in an age when people admired large populations, rather than fearing their effects.

It was a time of confidence, and warnings of danger were dismissed with a sneer, just as global warming is dismissed by some today. At year's end, the *New York Times* reported that Judge Clerke had administered 'one more rebuke to the much-abused Mr Malthus, and added the sanction of his name and position to the doctrine that the multiplication of our species is one of man's chief duties, and that any theories which propose to restrain him in the discharge of it are not simply erroneous, but impious'.

Malthus had died in 1838, but in the twenty years from 1839 to 1858, between his death and Darwin announcing his theory to the Linnean Society, the name of Malthus was invoked in *The Times* on 62 occasions, 31 of them in news stories, and 15 more in editorials and commentaries. The people of the 19th century were well aware of matters of population, but surpluses could still be easily shipped off to other available lands.

Those who claimed the Earth was young pointed to all the available space for spare humans as proof of their view. A fortnight before Darwin's book came out, there was a squabble in the pages of *The Times* concerning some flint tools found buried in chalk. In the course of this, a correspondent writing as Senex, a practised '6000-year' advocate, jeered at any suggestion that these stones were ancient. Without seeing them he asserted that these so-called implements were shaped by frost, or by pressure, or by being ejected from a volcano:

> Since the historical era our race has multiplied rapidly, but there are more regions still uninhabited than are occupied by man. Could this be so if man had existed on earth 14,000 years ago, as Messrs. Horner and Darwin pretend to show?

The supposedly uninhabited regions were filling fast. Canada's European population in 1859 was around 2.5 million, compared with 50,000 in New Zealand. Australia reached one million at the start of 1859, with 489,000 people in Victoria, 306,000 in New South Wales (which still included Queensland), 110,000 in South Australia, 82,500 in Tasmania and 14,000 in Western Australia. These three nations, the Cape Colony and the USA, were all drawing people away from Britain, and the British emigration trade was booming.

A full 7000 emigrants left Liverpool in September alone: 4800 to the US, 1000 to Victoria, 400 to New South Wales, 224 to New Zealand, 400 to the Cape (exclusive of assisted passages, more than 2500 Britons emigrated to the Cape of Good Hope in the first half of 1859) and a few to Canada. Nobody really considered the long-term effects, but each migrant arrived with a stock of methods, ideas, hopes, values, dreams and aspirations, all ready to be transplanted to a new home, another source of change.

The ships and their masters were controlled by the British *Passenger Act* of 1852 which laid down the conditions for the emigrant, and set a quota for numbers, requiring two tons displacement for each adult passenger.

Children counted as half an adult, if under 14 (or sometimes under 12), and those under 7 rated as one-third of a 'statute adult', while infants were free. Fares were calculated on the same basis, and provisions were also set out in the Act.

The *Annie Wilson* was chartered in July to carry emigrants to Sydney at £13/19/9 per statute adult. *The Times* advised that one of 'Dr Normandy's fresh water distilling apparatus and a person competent to operate it', plus fuel, would be supplied, allowing the owner to reduce by half the amount of water required to be carried under the *Passenger Act*. Normandy's device worked well, and his Patent Marine Fresh Water company flourished until 1912.

Fares were not cheap. In terms of purchasing power, the one-way fare on *Annie Wilson* would buy a return economy-class air ticket today, with a bit left over. John Wilson, his wife and three children travelled to New Zealand, arriving in early 1859 for £51, while a few cabin places were offered in January on the Aberdeen clipper *Dunrobin Castle*, about to set sail on 10 January for Sydney, at 30 guineas a head. *Wild Ranger* offered second-class berths at 20 guineas.

The fastest ship in the world, the celebrated *Lightning*, was due to leave on 5 January, and her owners claimed she had made Melbourne to Liverpool in 63 days, port to port, having once run, on 19 March 1857, '430 knots (sic) or 501 statute miles' in that single day. Passage money was listed in *The Times* as '£14 and upwards'.

When the government emigrant vessel *Escort* left Liverpool in late April, she carried 316 souls or 279 statute adults, said *The Times*. The group consisted of 41 married couples, 75 single men, 96 single women, 22 boys aged one to 12, 30 girls of the same age range, seven male and four female infants. A chief cabin to Halifax and Boston cost £22, a second cabin was £16. To New York, it was £26 and £18. By comparison, steerage passengers could travel from New York to California for $45 (about £9), bed and board found.

There was no shortage of takers: *Scientific American* reported that 1200 passengers left on the *Baltic* on a single day, and two other crowded steamers departed for California at the same time. By 1859, California had been a state for almost a decade; Oregon would become a state during the year. Gold was still being won in both, but Australia was proving even richer, and 1859 saw the first New Zealand gold finds at the Buller River.

The Comstock lode of silver ore was found in Nevada by Henry ('Old Pancake') Comstock (1820–70) and within a year, there were 10,000 people at Virginia City. By 1890, the mine had flooded and Virginia City was a ghost town.

Then there was the 'Kansas gold': the Pike's Peak gold strike report was made, then denied, then confirmed, the rush was on and 'Fifty-Niners' streamed into the Rocky Mountains. The *New York Times* reported in May that hundreds of men were arriving in St Louis, intent on reaching Pike's Peak. Yet, they were without equipment, tools, guns and ammunition or food—and stocks were fast running out.

The Australian goldfields had been stripped of most of the rich surface gold, but hard-rock mining was providing steady employment for many, and huge profits for shareholders. In 1858, between 27 October and 14 November, just under six tons of gold were shipped to Britain from Australia, and the flow continued unabated through 1859. The colonial venture was paying off handsomely for Britain!

Those supplying the prospectors always did well.

Individual miners continued to find gold, but many were still a little weak on their arithmetic. The solution came in Glass's *Gold Reckoner,* the work of Charles E. Glass, Bookseller and Printer at the Peoples' Library, Market Square, Castlemaine, but 'Sold by all Booksellers in Victoria'. It allowed the simpler digger to check the payment due for different weights of gold, reducing multiplication to a straightforward addition.

## THE INGENIOUS INVENTORS

Others on the Australian goldfields were both literate and thoughtful. A letter to *Scientific American* appeared in the 7 May issue. Mailed from Melbourne on 15 February, it was received in New York via Marseilles and London. In it, Messrs Fisher, Ricards and Co sought a pump able to clear water from deep mines, 200 to 300 feet deep. If some American could invent such a pump, they undertook to finance and manufacture it and share the profits.

*Scientific American* was enthusiastic about the expanding patent business, as might be expected of a journal whose proprietors were patent attorneys. They reminded their readers that when the US Patent Office had burned down in 1836, 10,000 patents were destroyed, the total production of 46 years. By 1859, the Patent Office issued that many patents in just 46 months.

Inventions of the year included things as different as a soda water bottle stopper, a mechanical clothes wringer, a new floor covering material, patented by Mr Dunn of Goodyear's India rubber company and described in the London *Mechanic's Magazine* as made of cork, flock, cotton, wool and other fibres, mixed with India rubber. This product lost out when Frederick Walton patented linoleum in 1860.

Then there was Beardsley's Ship Cooking Stove which was mounted on gimbals and fitted with neat sliding weights to adjust the balance when a partial load was on the stove. Every week saw clever, simple mechanical solutions that others could take and apply to new problems. It would be another 30 years before Mark Twain would create his fictional 'Connecticut Yankee at the Court of King Arthur', a time traveller who could make all his 19th century needs at Camelot, but the inventive spirit of the Connecticut Yankee was already alive and well, all over the world.

Inventor B. Frank Palmer was hailed as a master maker of artificial limbs, a cunning user of springs to prevent jarring and the thumping sounds so often heard from more basic 'wooden legs'. Within a few years, there would be a great need for such skill. Before World War II, amputation was the preferred treatment for a compound fracture of the leg, because infection was sure to set in where bone had pierced the skin, giving bacteria a ready made entry point. With anaesthetics available, and until antibiotics became common, a single clean surgical cut to remove the limb above the wound, neatly swabbed and bandaged, gave the patient a better chance of survival. New weapons, new explosives and bigger wars could be relied on to provide a steady supply of hideous wounds for surgeons to 'cure'.

The needs of war brought non-lethal advances as well. A French artillery officer called Amédée Mannheim invented the first slide rule, reducing the complex multiplication and division needed to aim guns to simple addition, just as Glass's *Gold Reckoner* had. Another Frenchman, one A.A. Boulanger, began selling a new type of binocular that used prisms in 1859. Called an 'astronomical double telescope', binoculars were patented at the end of November 1858, with an improved version in August 1859. The slide rule and binoculars would both provide more accurate artillery support, and serve in many gentler ways as well.

One might be excused for assuming from the name that the brassière was a French invention, but the first undergarment to resemble a bra was patented by American Henry Lesher on 17 May 1859. It was said to deliver 'a more symmetrical rotundity', but the only mention of support in the specification refers to providing a more graceful support to the skirts. It was a fairly impractical first attempt.

Equally impractical was the latest effort of Perry Davis of Providence, Rhode Island. He patented a buggy boat, a boat hull on wheels which could be hauled by a horse. It could be separated from the horse and had flaps on the wheel spokes which could be turned around to make paddle wheels of the buggy wheels, and a crank and gears to turn the wheels and carry the boat across deep water. Davis was described by *Scientific American* as 'the proprietor of the well-known "pain-killer"', a product mentioned by Mark Twain in *Tom Sawyer*. Tom calls it 'fire in a liquid form', and gets into trouble when he gives some of this 'most detestable medicine' to his aunt's cat. By the 1870s, the Pain Killer, patented in 1845, was distributed by Christian missionaries around the world, and even sold in Australia, so Davis probably made more money from it than from his boat buggy. Because 'Perry Davis Pain Killer' was a registered brand name, there was no legal requirement to list its ingredients on the bottle, but it was mostly opiates and ethanol.

If Mr Davis' horse proved restive hauling the buggy boat, it might be broken with Mr Bunting's 'Mechanical horse tamer', patented in London during 1859. This had two poles attached to a central post, heavy cartwheels, understraps and a trailing pole. The horse was harnessed into the contraption, the wheels were too heavy for it to buck, the understraps stopped it dropping, and the trailing pole stopped it going backwards. The horse was left with no choice but to behave.

There were theoretical inventions as well. Bernhard Riemann offered the Riemann Hypothesis in 1859. Among other things, this ties the average space between prime numbers near a certain number to the natural logarithm of that number. Around one million (with a natural log of 13), every 13th number on average is a prime, and at one billion (with a log of 21) one number in about 21 is a prime. Its accuracy and other deeper aspects of the hypothesis remain unproven.

## THE ENVIRONMENTAL COSTS OF PROGRESS

Some human activities came with a cost. Whale fisheries were failing, cod catches were diminishing, rabbits were released in Australia before the end of the year, and a number of species went extinct, while *Scientific American* was excited that hydraulic mining, the '... offspring of American genius' had been extended to Australia, Georgia and North Carolina. This used high-pressure water hoses, fed by gravity, to convert soil into mud and sediment that could be sorted for traces of gold. Its other by-product was in laying waste to the environment.

The US Patent Office distributed the seeds of cork oak and tea plants, in the hope that these economically useful plants might be grown in the US. Mulberry trees were being cultivated in Australia in the hope that a disease-free silk industry might be established, far from the silkworm cultures of Europe which had suffered from disease for half a century. North of Brisbane in the new Australian colony of Queensland, sugarcane was about to be farmed, and the colony's newly appointed Anglican bishop told an English audience that coolies would be imported to work it. He also regretted the difficulty in getting alpacas from Peru, apparently unaware of a flock of some 250 alpacas, already in Australia, which *The Times* said nobody seemed to want.

The demand for gutta percha meant that in Malaya, rather than tapping the trees, workers felled them and drained the sap, getting a quick profit, but endangering the species. The processes and wastes of industry, from chlorine works and paper mills to the leftovers of making coal oil and coal gas were all causing harm.

But did it matter, so long as things happened faster? In 1859, it was speed that counted—and materials and ideas about using them were spreading fast. New ideas, new methods and new knowledge all helped create the speed, and then rode on the new speediness, around the world.

# 2:
# The pursuit of speed

*Across the wires the electric message came,*
*He is no better, he is much the same.*

Former poet laureate, Alfred Austin (attributed)

It takes a very clever mind to predict the social impacts of an emerging technology. The first 20 years of a new technology are the domain of dreamers who simply want to make an idea work while everybody else shrieks that 'it' will never work. Old fogies mutter that the new technology will cause horses to break their necks, the young to rebel, the flowers to die, and the milk to sour, not to mention that it will encourage moral decay, tuberculosis, madness, and the end of civilisation as we know it.

At the end of 20 years, the early adopters are joined by the sharks who hope to seize control, but after 10 more years, the madness dies away and the new idea is grudgingly accepted. After 50 years have elapsed, the technology is taken for granted and its unexpected effects become clear. Railways, cinema, telephones, radio, television, air travel, space travel and the internet have all passed through (or are passing through) the same cycle.

When innovators make predictions about social impacts, they get the future wildly wrong, because they are too close to the idea. Marshal McLuhan compared them to drivers who look only in the rear-view mirror. We look to the past to name new inventions which is why we called cinema 'moving pictures', radio 'wireless telegraphy' and motor vehicles 'horseless carriages'—and we still speak of 'driving' a car, 'typing' on a computer and sending 'carbon copies' of emails.

In 1859, many new developments were helping to bring the world closer together, but only the keenest observers could see the changes they promised and threatened.

## TOYS FOR THE CURIOUS

Maurice Reynolds was a rare man. He saw the railway shrinking his city and his state, even if he never saw that the whole world was shrinking. In a public address at the Temperance Hall in Sydney in 1859, Reynolds looked at the railway, and lashed out at the experts. In 1859, two trains reached Sydney each day, one at 8.30 and another at 9.30 am. The afternoon trains left at 4.30 and 5.45 pm, running on a single track. It was an insufficient service in Reynolds' view; not a severe insufficiency, but he thought a couple of later trains would have been nice. So would some clear vision:

> The Commissioner [for Railways] himself, in a lecture delivered by him, maintained that Railways disturbed the cattle and annoyed the sheep, and Professor Pell in his little brochure broached a doctrine equally absurd, to the effect that fares ought to be fixed very high, otherwise every man, woman, and child would lose all their time and money running up and down the line continually.
>
> By means of railways the over-crowded and unhealthy parts of towns are enabled to scatter their inhabitants into the country; for the man of business can be as conveniently at his post from several miles off as from an adjoining street. In a word, such a change would be effected as if the country had been compressed by magic into a circle of a few miles diameter, yet without losing aught of its magnitude or beauties.

In short, the railway allowed healthy living, growth and suburbs. Reynolds missed noting that other new technologies of his time had similar effects.

The telegraph brought news faster than ships, injecting a new immediacy into the newspapers which were springing up to feed a growing literate population. The railways and the steamships brought overseas newspapers, books and mail to distant readers much faster, and people could travel more than they had done before, either in person or in their armchairs.

The travel book was once the domain of intrepid naturalists like Charles Darwin, Alfred Russel Wallace and Alexander von Humboldt. As tourism became the latest fad, ordinary folk began to publish accounts of their journeys.

Artists and scientists could travel more freely, sharing ideas and influences and collecting ideas. Their contributions and the growing riches in America, Britain and Australia, driven by gold discoveries in two of those countries, combined with Britain's booming industries, provided the capital and capacity for new projects that shrank the world.

Two major rail lines were completed in North America in 1859. One linked Baltimore through Louisville and Nashville to New Orleans. The other, the Montréal tubular bridge opened for traffic in December of that year. It cost $6,500,000, had 24 piers made of 3 million solid masonry blocks, and included tubes weighing 8000 tons. It was—for a while—the last link in Canada's 1000 mile long Grand Trunk Railroad, the longest continuous railroad in the world. Every civilised part of the globe had associated railway schemes. On 19 August 1859 Abraham Lincoln approached

General Grenville Dodge on the porch of Council Bluffs' Pacific House hotel; they talked for several hours about the potential route for a transcontinental railroad. In 1863, President Lincoln put Dodge in charge of building the Union Pacific line.

The many plans were not always immediately acted on. Though 8652 miles of new line was authorised during the railway mania of the mid-1840s Britain, the nation, had only 6621 miles of railway by 1850. This became 11,789 miles in 1875 and 15,195 miles in 1900. In 1845, there were 4780 miles of railroad in the US, by 1859 there were 28,238 miles. There were even dreams of a railway from Europe to Istanbul and on to India.

The first South African railway ran from Cape Town to Stellenbosch, Paarl and Wellington in 1859, and Australian tracks were being extended in and beyond the colonial capitals. The recently completed railway between Alexandria and Suez, some 200 miles long, had four locomotives: two English and two American. But despite the competition between the two nations, said *Scientific American*, the pasha, the local Ottoman ruler, had agreed to use American engines.

The journal was sometimes taken in by dubious reports. In December, its readers were told how locomotives were sometimes being fuelled by 'mummies'. These burned very hot, and the supply was inexhaustible: what a fate for a king, wrote the gullible reporter, who had not realised that there was a supply problem—even if the mummies were inexhaustible, royal mummies were not, and commoners must also be going into the fire-box.

Charles Dickens pointed out in *Dombey and Son* that MPs who had laughed at railways 20 years earlier, had woven them thoroughly into their lives by 1848. In 1834, a French minister, Adolphe Thiers, inspected the Liverpool to Manchester railway, and declared that 'railways were only toys for the curious, or a means of transport in exceptional cases only'.

Thiers was still active in 1859 when *Scientific American* reported that a single locomotive factory in Paris employed 1200 workmen, and France had 2624 locomotives. An official had just inspected the new line from Marseilles to Toulon and Toulon to Nice, and recommended using an extra thousand workers on the section near Esterel. A railroad linked Moscow and St Petersburg, a link to Warsaw from the Russian capital was completed in 1861, Warsaw was connected to Vienna, a branch line connected Silesia, which already had rail connections to Berlin. Networks emerged and timetables, once a tool of the military, began to rule people's lives.

Traffic jams were common in the streets of New York, but *Scientific American* thought the London police must be doing a better job of managing stoppages than New York's finest. Each day, 107,000 foot passengers and 20,498 vehicles crossed London Bridge without any jams, said the writer. Still, London had serious problems, and its congestion led to the idea of an underground railway to relieve pressure on surface transport.

The scheme was first proposed in 1837, but building did not begin until 1859. The first—the Metropolitan line—was very popular when it started running (at five to 10 minute intervals), and took a great deal of revenue from the horse omnibuses. But after the initial excitement, many customers returned to street travel. Even though the Metropolitan underground railways carried 25 million passengers a year, the London General Omnibus

Company still carried 40 million. Without trains, traffic would have locked solid, and people would have walked rather than be stuck on the bus, so the railways directly and indirectly benefited the omnibus proprietors.

Railway proprietors were subject to many curious ideas about this new technology. *Scientific American* said an 1859 correspondent of the *Daily Times* of Easton, Pennsylvania, who signed himself 'Inventor', had a cunning plan to keep the Pacific railroad clear of Indians, buffalo and other inconveniences. He would suspend the rails high in the air from balloons and hold the rails in place with magnets, buried in the earth at regular intervals. A telegraph wire would pass over the tops of the gasbags, and the lot would 'cost less than the common plan by some $400,000,000'!

By 1859, railways had been in existence for almost 30 years, beginning with the Liverpool–Manchester Railway in 1830, so it was an accepted technology. War demands speed, so it is hardly surprising that while fighting Austria in 1859, the French army moved some 600,000 men and 130,000 horses by train.

In July 1861, the first battle of Manassas (or Bull Run) in the US Civil War was fought over the control of a railroad junction, and Confederate reinforcements arrived by train.

## THE RAILWAYS MATURE

In 1832, English eccentric and author, George Henry Borrow, walked from Norwich to London, a distance of 112 miles, in 27 hours.

According to his biographer, Herbert Jenkins, Borrow went to meet with the secretary of the Bible Society, and during this trip, he spent fivepence-halfpennny on milk, bread and apples. People back then had to walk if they could not afford a horse or a coach. Some were even paid to walk or run: footmen ran beside their masters' coaches, guarding the coaches against being overturned.

A running footman could complete 100 kilometres in a day, and keep it up, day after day. In 1773, Foster Powell was a 'pedestrian', famous for walking long distances for tiny bets. He walked the 650 kilometres from London to York and back in five days and 15¼ hours and later did it twice more. In the first half of the 19th century, tourism was often on foot, even for the rich, but it went a great deal slower than a running footman's pace. Even policemen walked and after P.C. Lewis Jones joined the force at Gorseinon in Wales in 1859, he noted a colleague's 'recipe for curing swetty feet':

> One ounce of salts desolved in a pint of boyling water, then add the quantity of gin, for to make it pleasant to drink, then drink a wine glass full when required.

In 1775, Dr Samuel Johnson and James Boswell went on a walking tour of Scotland and the Inner Hebrides. Boswell's famous account of the tour, recounts their use of local unscheduled boats and carts. In 1851, Wilkie Collins' *Rambles Beyond Railways; or, Notes in Cornwall Taken A-Foot*, shows us that rather than ending walking tours, the railways extended and regimented them. Johnson and Boswell set their own pace; but trains and steamships ran to timetables, imposing a new urgency on people using them, even ramblers.

Stage coaches started it, but timetables were now everywhere. In cities and towns, steam whistles, bells, flags and lights on towers indicated impending or recent arrivals and departures, summoning those to start walking who planned to meet (or be) passengers, and those seeking to send or collect goods or mail.

Railways were the height of wonder for their time, shaping the era and the way its populace travelled and mixed, but even they would be supplanted by newer and grander modes of transport. Indeed the Suez railway was soon replaced by the canal that was started in 1859, and the world's first trans-continental railway at Panama, completed in 1855, had prompted adventurers to explore places where a canal might cross the isthmus, but for now, rail was the best for rapid travel. In its absence, steamers were useful as well, and coordination through timetables became common.

Edward White, a Canadian Wesleyan missionary, crossed Canada the quick way, via New York and Panama. He left Toronto on 1 January, departed New York by steamer on 6 January, reached Panama on 14 January, took the train across and reached San Francisco by steamer on 29 January, arriving in British Columbia on 10 February. He recorded the Panama crossing in his diary:

> About 10 a.m. we saw the coast of South America. Beautiful beyond description. Land at Aspinwall at 1½ p.m. Started on rail at 3 and got to Panama at 6. Waited two or three hours for tide and got on the John L. Stephens 9 p.m. A beautiful day. Enjoyed the ride across the Isthmus. Much of the scenery magnificent.
>
> Edward White, Diary, 1859.

Conditions on the trains were variable. Many lacked heating or lighting, and passengers needed their own candles, though by 1859, lighting was getting better. Steam heating came in 1874, dining cars were introduced in 1879, and on-train toilets first appeared in 1892. Until then, gentlemen could buy long tubes that strapped along the leg, while ladies might travel in all-female groups and bring a chamber-pot, hidden in a basket. An open window might bring more than a smut to lodge in one's eye, and it might not have done to drop things on the floor, either.

Eccentrics loved trains. One conceived the 'sound gatherer' to help a train driver hear what was on the line ahead. A huge ear trumpet was to be placed at the front of the locomotive to funnel all the noises back to the driver, no matter how much clatter the train made.

Real disasters and the fears of potential travellers also inspired inventors. The Loughridge patent brake of 1859 could stop a train going at 30 mph, in 600 feet, 'without damage to the engine or cars'.

Improvements were certainly needed. At Schaghticoke near Albany, New York, a bridge fell down under a passing train: nine people were killed and 17 wounded, some dangerously. The 60-foot wood bridge of Tomhannock Creek was found by a coroner's jury to not only be rotten but known to be so, and heavy damages were expected, said *Scientific American*. The railroad company collapsed later under the pressure of lawsuits.

In Cuba, a draconian way of encouraging safety had been put in place. Under the heading 'Rather Tough', *Scientific American* explained: 'It is reported that Mr Perry Bigly, formerly a citizen of Macon, Georgia, has lately been garrotted in the city of Havana, on a charge of carelessness or negligence in running a train, of which he was the engineer, by which a "run-off" occurred and several persons lost their lives'.

By 1859, most railway systems were being directed by practical civil engineers. A test train ran 412 miles on two and a half cords of wood, another travelled a mile on 18 pounds of anthracite coal. All the same, newspapers reported, scathingly, that the New York and Erie railroad was bankrupt, adding that the failure was due to nepotism on the part of the directors, which led directly to bad management.

One of many sleeping car designs for US railroads around 1859.

The first-ever sleeping car was built for the Prince of Wales to use in 1860 when he went to open Montréal's Victoria Bridge. George Pullman saw it and obtained a US patent for a sleeping car in 1859. Today, 'Pullman' is synonymous with sleeping cars, but Pullman's first version only appeared in 1863, while other designs had been around since at least 1857. One design described in *Scientific American* in 1859 drew the comment that 'if there were a line of railroad to the Pacific, an army might be transported perfectly fresh in a few days, by one continual trip from New York to California'.

In May, the *New York Times* told its readers that passengers who left New York by steamer on 9 March arrived in San Francisco 28 days later, on 5 April. Steamers, said the paper (two steamships and one train), usually made the trip in 21 days to 23 days.

## ACROSS THE WIRES

In 1871, the world held its collective breath as Queen Victoria's eldest son, Edward, Prince of Wales, was stricken by typhoid, the disease which 10 years earlier had killed his father. E.F. Benson claimed later that the po-faced poet laureate of the day, Alfred Austin, penned the versicle which appears at the head of this chapter.

Benson is the only source for this story, but he made the dry giveaway comment that it *sounded like Austin*. It does indeed sound like him, but it almost certainly wasn't. Still, where Queen Victoria was the exclusive recipient of the 1859 news of her grandson's birth, the 1871 news of the Prince's trials and eventual recovery could be wired to all Britons, Americans and Europeans and much of Asia. A year or so later, even Australia—on the other side of the planet—was linked in.

International telegraphy began in 1850, when a cable was laid across the English channel by John and Jacob Brett. Legend has it that a fisherman found the line and cut it, thinking it was a new kind of seaweed. This tale apparently originated with journalist and war correspondent, W.H. Russell, but neither Brett brother ever mentioned it, so it may be a journalist's yarn, or relate to some later event. In any case, the line failed, but the Bretts made a second link between France and Britain in September 1851. This time the cable worked.

Paul Reuter set up a carrier pigeon service in 1850 between Brussels and Aachen, linking French–Belgian and German telegraph systems. Then in 1851, he set up a telegraphic news agency, using the Brett cable, augmented by 200 carrier pigeons. In 1852, Chile, Santiago and Valparaiso were linked by telegraph.

The decade of the 1850s saw massive telegraph expansion on land. In 1845, there were 900 miles of telegraph line in the US, by early 1859 there were 30,000 miles. As early as 1852, the first issue of the *American Telegraph Magazine* forecast future lines to Europe, China, and even Australia. By 1857, most large towns in England were linked by telegraph.

In 1859, lines and cables were being erected, laid and used all around the world. The first complete Atlantic cable worked briefly in 1858 before failing, and several others were proposed.

There were 19 different under-sea cables laid in different parts of the world during the year. Among others, Malta was linked to Sicily, Sweden to Gotland, England to Heligoland, Singapore to Batavia, the Australian mainland to Tasmania. Landlines too, were increasing.

After an unsuccessful first attempt in 1857, the second Atlantic cable was laid in 1858. It worked briefly, but failed after three weeks, in which time it had carried some 400 messages, one of them an exchange between Queen Victoria and President Buchanan. There would be no permanent connection until 1866, but the idea, the ideal, was there.

The 1858 Atlantic telegraph was ponderously slow before it failed, but Professor Thomson, the future Lord Kelvin, argued that even one word a minute could be of amazing value. The cable had twice as much metal and gutta percha as any previous cable, yet the induction effect—predicted by Michael Faraday before the cable was laid—had slowed the signals enormously. Much faster transmission was what the world wanted.

## Extending the cables

Making submarine cables was a challenge. The conductor needed to be as large as possible, to keep resistance down, the insulation had to be impenetrable, and in the days before rubber could be well managed, long before plastics, the only reliable insulator was gutta percha, which still had its weaknesses.

The Australian line under the sea to Tasmania opened in September, but it soon failed when waves at Cape Otway on the

Victorian end dashed the cable against rocks. The cable needed armour to save it from sharp rocks, but adding extra material carried a cost. The armour made it heavier; when a ship lowered cable to the ocean floor, there could be several miles of cable dangling at any time, and heavier cable was more likely to break under its own weight. Also, a heavy cable was unlikely to be recovered for repairs once it was laid. A fault in a landline could be fixed, but a fault under the sea usually meant a dead loss.

Ideas for 'better cables' were common. D.E. Hughes of New York offered a self-sealing cable with the wires in a gutta percha tube, surrounded by a semi-liquid ooze of either rosin dissolved in oil, or India-rubber dissolved in naphtha, or rosin soap. If the cable were to be breached, the semi-fluid would seep out and harden in the seawater, closing the break. *The Times* reported a new form of submarine telegraph cable developed by Messrs Wells and Hall, using pure India rubber, rather than gutta percha, whose coverings were braided to prevent kinking.

Other manufacturers used pitch, linseed oil, hemp and tar in varying combinations. A new submarine cable had been laid from England to France, this one weighing a massive 10 tons per mile, 'the largest and strongest cable ever made'. It contained six wires, each as large as the whole number in the Atlantic cable.

Telegraphy was a business, and it could be a profitable one, but some of the directors needed to study the economic issues more closely. The conductor in a mile of early submarine cable was copper weighing some 63 pounds, valued at about 63 shillings, just over £3. Wrapping gutta percha around the conductor added £40 a mile, and the whole cable, with iron and gutta percha sheathing and copper, cost £100 a mile before it was on board ship. Doubling the copper increased the cost per mile by about £3, so there was a clear economic case for spending a little more and using much more copper to reduce induction and give much faster signals. But it took time for the money men to accept that reducing the amount of copper was a false economy.

In August, the Electric and International Telegraph Company, owner of some cables linking Britain to Europe, held its annual meeting and noted that as cables became older, failures would inevitably be encountered, and these must be planned for. Their engineer warned that over time, gutta percha decayed, and there was no known way of preserving it.

Even where grand schemes for international networks were not in place, short pieces were being assembled. One such was the England–Australia connection. With the three main Australian colonial capitals, Sydney, Melbourne and Adelaide linked by telegraph wires strung on poles, a connection to any one of them would tie in all three, but the signals would travel through many countries on many separately installed elements before reaching their destination.

Most of the sections from England to India were in place or planned, but there remained the undefined link from Asia to somewhere in Australia. From India, messages would pass through Ceylon (Sri Lanka) to Singapore, and on to Batavia (Jakarta) but then came the last hurdle.

There could be overland cable on Java and a submarine cable from eastern Java to Moreton Bay (Brisbane), then overland; a submarine cable to Perth, followed by an overland route along the southern coast or a submarine cable to Port Darwin, then overland to either Brisbane or Adelaide.

An English businessman, Lionel Gisborne, had joined an American businessman, Cyrus Field to lay the 1858 Atlantic cable. Building on this start, Gisborne raised capital for the Red Sea and India Telegraph Co, proposing to link India and Australia in the slightly longer term. He sent his brother Francis to Australia in July and the news of this mission inspired a comment in the *Gentleman's Magazine* that a cable to Australia would soon be a reality. The brutal reality was that Francis could not get the colonies to agree about either a route or payment.

The jealous colonial powerbrokers struggled for control of the vital commercial intelligence the telegraph would carry. In 1859, the South Australian government offered £2000 to the first person to cross the continent and reach the north coast from Adelaide. After a number of attempts, John McDouall Stuart reached the coast in the middle of 1862, and desperately ill, arrived back in Adelaide at the end of the year. He died in June 1866, never having seen any more than the interest on the £2000.

During 1859, the burghers of Melbourne, to the south of Sydney and the east of Adelaide, conceived a scheme to hijack the fast communication, to find a route to Melbourne, all the way from the north, a route far enough from both Sydney and Adelaide to avoid any direct connection to either. The links from India to northern Australia were being prepared, so there was no time to waste if Melbourne wanted to seize control!

The Melbourne organisers sent for camels in 1859, and in 1860, the ill-prepared and ill-fated 'Burke and Wills' expedition set out from Melbourne aiming to reach the Gulf of Carpentaria in Australia's tropical north. They reached the north coast, but Burke and Wills both died on the way back. It would take some ten years before a line was built, and between 1870 and 1872, the Overland Telegraph was built along Stuart's route to Adelaide. The Melbourne planning and plotting had achieved nothing but death.

Everywhere, transmission speeds were improving. British scientist and inventor, Professor Wheatstone's automatic telegraph could print 50,000 letters in a minute, using a perforated strip of paper to 'receive records'.

This was an intermediate stage before transmission, after which a recording machine at the other end placed marks on a second strip, ready for somebody to read. Critics, who were drawn to every new invention like vultures to carrion, said the idea was derived from older technology. Perhaps it was, but nobody had combined them so neatly!

## The telegraph in wartime

The telegraph interested military minds. A detachment of the Royal Engineers left England for China in November. A number of photographers were going, so were telegraphers with 'several miles' of wire, a most inadequate ration, even then. In mid-1859, the *London Globe* had reported on French military telegraphy: from each corps, a horseman would ride off to the next division, unreeling a light wire, until a front of twelve miles was connected by telegraph. Other wires, behind the front, connected the army to Paris and Piedmont. The Austrian army, said the paper, had demonstrated just such a flying telegraph several years earlier.

The world's armies began to rely on the telegraph during the Crimean War, when the British and French governments wanted to communicate directly with commanders at the front. Unexpectedly, the line also allowed *The Times* to deliver news of what was happening, almost as it happened—and that would change warfare forever. At the start of the war in 1854, a message took two days to get from the Crimea to Varna by steamer, then it went three days by horse to the nearest telegraph station at Bucharest. The Ottoman empire's pashas, the semi-independent local governors, had opposed any telegraph on their territory citing religious prohibitions. Basically, they were worried about interference from Istanbul. But when it came to having to fight a war, their objections dissolved—Islamic tradition allowed innovations that might win a war. So they lost their case and the cabling began.

By October 1855, Istanbul and Europe were linked, but Britain wanted a line to India, and even before the Indian troubles of 1857, the British ambassador to Istanbul, Stratford de Redcliffe, had started negotiations to run cables through the Ottoman empire.

There was, however, an unexpected consequence. Even when the operators were Turkish citizens who used Arabic script, civilian traffic was in French, and later in English—the telegraph used the Roman alphabet. So, younger Turks had more reason to learn foreign languages. As had the Chinese and Japanese, the Ottomans saw the telegraph as a way for the West to infiltrate and dominate.

Looked at more neutrally, the telegraph gave people of different cultures a common cultural space to exchange ideas, it brought separated parts of the world closer together. In short (no pun intended), the telegraph shrank the world.

Work on the cable between Britain and India began in 1859, but it took another seven years to complete a successful and lasting telegraph link all the way to the subcontinent.

In 1857, news of the 'Indian Mutiny' took between 30 and 50 days to reach London, but the internal telegraph system linked upper India with Calcutta, and so to Madras and Bombay, letting local British troops coordinate their responses. 'It is that accursed string that strangles us!', one of the Indian patriots is said to have complained, glaring at the telegraph wires as he went to his execution. By May 1859, with Suez–Aden complete, a message could go from Bombay to London in just 10 days, but soon after, the Red Sea cable failed. The contract had foolishly allowed the

The telegraph office, Western Union Company, New York.

company laying it to keep any cable left over. They scrimped, the line was laid with little slack, and before long, it broke.

Messages travelled as best they could. A month after the execution of one of the last 'Indian mutineers', *The Times* was able to report on 19 May 'from Malta by submarine and British telegraph: Tantia Topee was hanged at Seepree on the 18th of April.' By hook or by crook, by horse or by cable, the speed of communication was picking up.

Work on a line between St Louis, Missouri and San Francisco began in 1859, with horse riders and written messages spanning the shrinking gap between telegraph stations. It was completed just after the start of hostilities in the US Civil War. For its effects in uniting the United States, this nation-shrinking link was even more important to the USA than successful telegraphic communication with Europe.

The 40 miles of line laid from Washington to Baltimore in 1844 had become 40,000 miles of wire in 1859; telegraph operators were also getting much faster. The first telegraphers were expected to read the tape from a recording device and dictate the message to clerks before a third clerk made a 'fair copy'. The recording device made clicks and operators found they could decode 2000 words an hour without looking. By 1859, every telegraphist hired had to be able to receive 'on the click'.

In early 1860, a single telegraph operator sent 2083 words by Morse in one hour between Philadelphia and Pittsburgh. In one day, operators sent 578 private dispatches, 5000 words of news for the Associated Press, and an entire copy of the President's message, more than 15,000 words, for the *Pittsburgh Post*. Two wires were used, and it took 5¼ hours but it was still within a single continent.

Still on the search for a way to cross the Atlantic without problems, the Canadians had another idea to join North America and Europe: go the other way! A bill was passed by the Canadian parliament in mid-1859 for a line across Canada, over 'Behring's straits by a short submarine cable' then through Russian territories to northern Europe. It would go from Russia to Victoria and then branch, one arm to San Francisco and one across Canada. From San Francisco, it was 5000 miles to Novgorod, and the line to Missouri was already under way in the other direction. The plan was taken seriously in 1859–60, and was only dropped in 1867, when Cyrus Field completed a successful and robust new Atlantic cable. But there was a curious aftermath.

Some of those who worked on the construction in what was then Russian America were the same people who advised William Seward and Charles Sumner when they persuaded the US Senate to purchase today's Alaska. One can never predict the effects of a technology!

## GIVE US THIS DAY OUR DAILY NEWS

The speed of communication was a marvel. *Scientific American* reported in September that a message had gone from Albany to Kansas City, and a reply had been received, all in six hours. In Britain, speeches by Mr Bright and Mr Gibson in Manchester on a

Friday night began being sent to *The Times* at 10.55 pm, and by 2.45 am on Saturday, a report occupying almost six columns, was set and appeared in the Saturday morning paper. Now the news in a newspaper really *was* new!

The telegraph added an immediacy to newspapers, which now had to be up to date. The growing literate class in the English-speaking world was served by a variety of weeklies, from *The Scientific American* to *The Illustrated London News*, and more and more dailies, all driven by a sense of urgency, a hunger for novelty.

In Britain, the growth was assisted by the repeal of various taxes on newspapers. From 1815 to 1836, the stamp tax on newspapers was fourpence. The advertisement duty was reduced in 1833 and repealed in 1853, the stamp duty was reduced in 1836 and repealed in 1855. The duty on paper itself was abolished in 1861.

The *Irish Times* and *Munster Express* were both started in 1859, but growth in the newspaper industry was almost as healthy in the US where there were no taxes on paper or papers. By 1859, there were 4000 US newspapers, 500 of them daily, and 500 semi-weekly with a total circulation estimated at 400 million issues a year, against about 20 million in 1813.

Competition heated up everywhere. In London, *Bell's Life* was pirated in 1859 by a paper calling itself the *Penny Bell's Life*, and the proprietors of the *London Journal* took out a successful injunction against the *Daily London Journal* which was planned by the person from whom they had bought the *London Journal*, after he had undertaken not to publish any weekly journal of a similar nature. The lawyers had plenty to eat that month.

## The presses get faster

The presses also 'ate', because type metal was soft, and soon wore away when it was used to print directly on paper. In the days of hand-setting, each letter was selected, one at a time, to assemble a line of print which was then placed in a frame, below the earlier lines, composing a column or a page. It was slow work, and if letters were worn from printing, they had to be melted down and new ones cast, one letter at a time.

By 1829, papier mâché stereotypes were commonly used by printers to make plates from which they would print. In 1859, *Scientific American* readers were advised that their magazine was now printed from the copper plates of the electrotyper, and that even as one issue was being printed, the type was being broken up and sorted, ready to be used in setting up the next issue.

In the electrotyping process, graphite (also called plumbago or black lead) had two separate uses. First, it provided a slippery surface that wax would not attach to, and second, it formed an electrically conductive surface to which electrolytic copper would attach when the plate was being made. The process began with normal hand-typesetting. The type was locked solid and wiped over with graphite lubricant before hot wax was pressed down on it in a hydraulic press at 1000 pounds per square inch. This formed a mould, a negative copy of the type, after which the individual letters could be separated ('broken up') and returned to the compositors.

Those parts of the mould where copper was not to be deposited were coated with wax, then graphite was brushed over the rest of the mould to make it conductive. A battery was used to plate the surface with a thin copper shell, an exact copy of the type. The shell was removed, tinned on the back, and filled with molten type metal to make a solid copper-faced block, just like the original hand-set type. This was the printer's plate.

There was also a method using gutta percha for the mould, and some printers preferred papier mâché flongs that could be prepared from flat type and bent into a curve to form the cylindrical plate needed for a rotary press—and as steam presses came to be used more, the rotary press became the standard form. An American, Richard Hoe, had invented the steam press in 1843, taking out a patent in 1846. By 1859, *Scientific American* could announce proudly that a six-cylinder printing press had just been shipped from New York to Sydney, and that Messrs Hoe and Co had sent out 'one of their best men to superintend its erection'. The technology of speed was being exported around the globe.

Novelists were, in a sense, a product of the steam press—and the steam locomotive that could distribute their products. The steam press had to keep running if printers were to recoup their investments in equipment, so printers began to diversify. One way was to publish novels, though it was also capital-intensive, tying money up in type and stock. So, novels often appeared first in Britain as 'three-deckers', triple-volume releases that were sold mainly to private circulating libraries. The railway also provided new outlets and a new demand.

By 1859, rail passengers needed reading material and the railway bookstalls were a major source, as this advertisement from *The Times* illustrates:

> NOTICE.—BOTANY BAY, by JOHN LANG, Esq.—The publisher having received numerous letters from gentlemen not being able to procure this work at the railway book stands, Mr. TEGG begs to say that it can be forwarded free by post for 3s. 6d.

The cheaper 'railway editions' of these releases sold on the ubiquitous bookstalls of William Henry Smith, father and son, who began trading at Euston Station in 1848. The younger Smith later went into politics, became First Lord of the Admiralty, and was lampooned by W.S. Gilbert in *HMS Pinafore*.

## THE MAILS GO FASTER

With railways spreading fast and publishing booming, the postal services grew. From the 1840s onwards, starting with the Penny Post in Britain, radicals and thinkers had the post as a new tool for organising. Adventurers seeking gold in America and Australia, emigrants and others, all boosted the mails, but the mails carried ideas as well as news.

It was an age of letter writing. Just before he died, Alexander von Humboldt was reportedly answering 2500 of the letters he received each year. Pillar boxes had been introduced to Britain in the 1850s, they began to appear on Philadelphia streets in 1859 and New York acquired some later in the year. Before that, US mail collectors would call at various buildings to pick up the mail left in tin boxes.

Early in 1859, the Peninsular and Oriental Steam Navigation Company won the contract to provide regular mail services to Australia via Suez. The first P&O mail steamer left Southampton for Alexandria via Gibraltar on 14 March, carrying with it the parcels that were designated 'heavy mail'. Lighter mail left London on 18 March, crossing France by train, leaving Marseilles on 20 March by steamer for Alexandria, where all the mail was railed to Suez. From there, a steamer took the mails to Mauritius where 24 hours was allowed for coaling before the ship sailed for King George's Sound, where the Perth bags were dropped, with another 24 hours allowed for the ship to take on coal.

Then it was on to Kangaroo Island where the mail for South Australia was passed to a branch steamer from Adelaide (two hours allowed) then to Port Phillip heads, where the Geelong mails went ashore and the Sydney branch steamer took the New South Wales mails. Then the ship stopped in Hobson's Bay to drop the Melbourne, Hobart and New Zealand mails, with branch steamers going to Hobart and New Zealand. The time allowed: 52 days from Southampton, 45 days via Marseilles. Six hours after Kangaroo Island, the first mails reached Adelaide, and the main news could be telegraphed to the other colonies.

## The mails become a business

Then came a shock: the Australian colonies were appalled to learn that the British Post Office planned to double the postage on newspapers. A deputation from the colonies met with the Postmaster-General, Lord Colchester, to express their outrage that the increase should be proposed so soon after a mail contract had been let, noting that newspapers—at a penny—were cheaper than letters, and were a preferred way among the poorer classes for family in England to maintain contact.

The Australians argued that now the mails crossed Suez by train rather than by camel, the price ought to have dropped. They noted that colonial newspapers were carried to Britain free of charge from Tasmania and New South Wales, and papers could be carried from one part of England to another for a penny. Given that a member of the government had recently called Australia 'an integral part of England', the same price should apply. Colchester said he would postpone the rise until 1 January 1860, but argued that the same charge applied to India, just as much an integral part of England in his view.

The mails were big business all over the world. On just one day, Wednesday 30 March 1859, the New York Post Office dispatched 35,187 foreign letters, while three steamers, *Niagara*, *Kangaroo* and *Persia*, brought 72,499 articles from overseas. A further 90,000 domestic articles were sent and received.

By 1859, Dickens' magazine *All the Year Round* could offer an account of the railway Post Office, where mail sorting took place on the train, and bags were collected and delivered on the fly—deliveries bagged and held out to be snatched from spring-loaded hooks while new mailbags were taken up in nets.

Travellers, however, were treated more gently as they joined and left the train. Now people could board a train and go undreamt-of distances in a single day, and see wonders they had only ever expected to dream about. The Age of the Tourist was dawning.

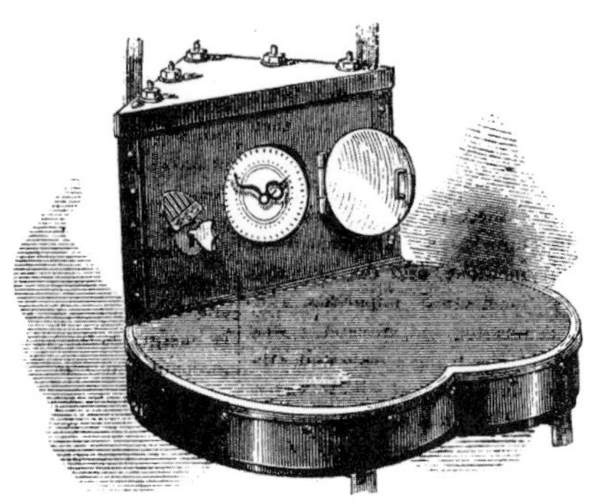

# 3: Faster travel

*I will bet twenty thousand pounds against anyone who wishes that I will make the tour of the world in eighty days or less; in nineteen hundred and twenty hours, or a hundred and fifteen thousand two hundred minutes.*

Phileas Fogg, Jules Verne's *Around the World in Eighty Days*, 1873.

The tomb of Henri Mouhot sits just above the Nam Khan River, not far from Luang Prabang in Laos. He died of malaria in 1861, while on his fourth collecting expedition in southeast Asia.

Covered by jungle until 1990, the tomb is visited by a handful of scientists each month, people who ride out on the road (dusty or muddy, depending on the season), walking down to the river and along its litter-strewn shores, before stepping up into the jungle again.

Sometimes, domesticated elephants walk through the area, but Mouhot would have liked the rich plant and invertebrate life surrounding the tomb that was built to mark his resting place.

Yet while he sought fame as a collector of exotic plants and animals, Mouhot's lasting effect on Asia came from his 'discovery' of Angkor Wat in 1859; to be more accurate, his influence came from his posthumous book which drew European attention to Angkor Wat.

His tomb receives sparing attention from specialists, but tour mobs mill around Angkor Wat in monoglot coachloads, following their guides like so many terrified ducklings after their mother, scrambling on the carvings, surreptitiously rubbing their sweaty hands over the breasts and buttocks of the buxom stone Apsaras, looming rudely over the few Buddhist monks, screaming with excitement. This monumental ugliness is the curse we call Tourism.

## THE BIRTH OF THE TOURIST

Travelling from London to Rome took 21 weary days in 1843. It took just two-and-a-half days in 1860, and tourism had become a mass commodity. Just as the new technology brought fast travel to the masses, so Thomas Cook brought the masses to fast travel. He augmented the technology with group excursions, travellers' cheques, hotel coupons and round-trip tickets, teaching first a nation, then the world, to obey timetables.

Cook began simply, arranging for a temperance group of 485 people to go from Leicester to Loughborough in 1841 at a shilling a head, with a brass band, speeches and food thrown in. He built up slowly, then left his job in 1845 to run tours full time. By 1848, he was taking parties to Scotland and the Lake District, then running 'specials' to see the Great Exhibition in 1851, and private trains to London to see the Duke of Wellington lying in state before his funeral in 1852.

Mr Cook's train with 28 carriages of paying customers ran across Brunel's magnificent Saltash Bridge linking Devon and Cornwall, when it opened in 1859. Ten years later, he was offering a 105-day tour of Egypt and Palestine, and one of his 1872 brochures is reported to have inspired Jules Verne to write *Around the World in Eighty Days*. Verne seems to be one of the few writers of his time who was sensitive to how the world was shrinking.

Cook offered a world tour lasting a more relaxed 222 days but Verne's Phileas Fogg used timetables to find a faster way. In 1859, it might have been possible for a New Zealander to reach London comfortably in eighty days, via Suez and Marseilles, but getting right around the world in that time was still a bit of a challenge.

Some travellers had more pressing reasons to travel than 'merely' seeing the world—like dodging spouses or obligations. The classifieds in London's *Times* of 3 January reveal that people could advertise for missing friends in Australia; a New York enquiry agency offered to provide information about traders or to collect debts. In London, Charles Frederick Field, a former Chief Inspector of the Detective Police of the Metropolis, offered London and Continental Private Inquiries, and access to his New York agent.

Just before Christmas 1859, the *New York Times* reported that a burglar named Hod Annis had been taken back to Boston. Annis had been arrested in Philadelphia, and managed to escape the clutches of the law, but he was undone when he telegraphed his mistress to send him money, and the authorities intercepted his requests. His travels had been in vain.

Before about 1859, travel books were written either by intrepid explorers to recoup their costs, or as guides for intending emigrants. With steamships making foreign travel more available, the need for guidebooks grew—*Murray's Guide to Madras and Bombay Presidencies for 1859* was the beginning of a long line of them. The author was Edward Eastwick and the 'Murray' who usually gets more credit than Eastwick was John Murray, the publisher of Charles Darwin's *On the Origin of Species.*

Eastwick advised his readers that on arriving at Madras (Chennai), the trick was 'to get into a palankeen and be carried to the club, if a bachelor; or if travelling with ladies, to some friend's house. There are, indeed, hotels which may be repaired to as a dernier ressort'.

Isabella Bird Bishop, who became the first woman member of the Royal Geographic Society in 1892, published her book *The Englishwoman in America* in 1856. She was publishing still in 1900 having started when her clergyman father sent her to America to research American Christianity for him. A relative of William Wilberforce and cousin to John Bird Sumner, Archbishop of Canterbury from 1848 to 1862, she was well enough connected to get an introduction to John Murray, who accepted her first work. She never looked back.

After the death of her parents, Bird Bishop sailed for Melbourne in 1872, then visited New Zealand and Hawaii before crossing America and reaching New York at the end of 1873. There were more travels, and she circumnavigated the world three times, a record that few humans could have matched in the 19th century.

And it was all down to steam transport. Steamships were also used to take an England cricket team to North America in 1859, the first international tour involving any team sport. England won every match.

Sometimes, travel failed to broaden the mind. An American visitor to Berne had returned claiming that everything in Berne smelled of cheese. 'Cheese is the Bernese otto of roses', he grizzled, using a then common form of 'attar of roses', an oil distilled from rose petals, and still used in many perfumes. Every city had its smells, but the odour of horse dung was common to all of them.

## TRAVEL AROUND TOWN

The horse-drawn omnibus was introduced into England in 1829 by George Shillibeer. The earliest omnibus was pulled by three horses and carried 22 people, all inside. A smaller version used in 1849 carried 12 passengers inside and two outside. Horses were good for seven or eight years in the middle of the century but by 1894, a horse was worn out and ready for the knacker's yard after just four years on the buses.

Inventors rose to the challenge of packing in more people, especially targeting women in fashionable crinolines, and *Scientific American* reported that Mr Singer, the sewing machine inventor, had constructed a carriage which held 12 people inside and another 14 outside. It had a baggage compartment, a water closet, a room for ladies to arrange their toilet, a place for carrying dogs and poultry, with a total cost of $3000. It was to be drawn by six horses.

That rather pales into insignificance when compared with a legendary Australian coach, the 'Leviathan', which by various accounts carried 60, 76 or 89 passengers, drawn by either 16 or 22 horses. My use of 'legendary' is deliberate, because it may not have been built, though the tales come complete with varying measurements and the name of the driver, 'Cabbage Tree Ned' Devine, who allegedly drove it on the Geelong–Ballarat route from 1859 to 1863, or possibly on the Castlemaine to Kyneton route. Devine did, in fact, drive a large coach with 12 horses for a touring

England cricket team in 1862, which cuts across the alleged dates for 'Leviathan' to be in his charge. Still, giant coaches on the omnibus pattern were at least plausible.

In 1859 there were four different horse-drawn omnibus routes in the Dublin area, while Caroline Healey Dall, reformer, feminist and diarist, rode an omnibus in Concord Massachusetts in 1859 while on her way to visit Mrs Alcott (the mother of author Louisa May Alcott) before she gave a lecture that night. In 1836, *The Times* had seen fit to note that passengers should not spit on the straw, and that dogs should be on a short string. By 1859, Parisian omnibuses were admired because the floors had no matting or straw, and there were gaps between the slats, through which dirt could fall and fresh air could enter.

Dust was a problem on some roads, but a Lyons chemist offered a solution: sprinkle the streets with muriatic acid! *Scientific American* expressed concern about the effects on horses or pedestrians, because the writer thought they might be burned by the acid, prompting a reader to reply that weak hydrochloric acid was readily available and of no great use: if the streets were limestone, this would form calcium chloride, which absorbs water from the atmosphere because it is hygroscopic. Until the next rain washed it away, the calcium chloride would keep the road surface damp, and so help to keep the dust down.

In some cities the streets were being taken over by 'horse railroads', horse-drawn trams running on rails flush with the road. The Boston horse railroad carried 8 million passengers a year. Moving that many by omnibuses or carriages would undoubtedly block the roads, said the experts. Liverpool in England and Chicago tried horse trams in 1859, and Baltimore and Pittsburgh had them by the end of the year. The Chicago horse cars carried 20 passengers on benches, pulled by one or two horses. *Scientific American* told its readers a horse could haul five times as many passengers on rails, as it could over cobbles.

Hackney carriages of various sorts—also called cabs or hacks —were also used for transport. However, there was a snag. These cabs often doubled as hearses, ambulances or even prison vans, and a debate erupted in the pages of *The Times* late in 1859 when a number of letter writers claimed street cabs were being used to transport 'filthy and infectious persons'. At the time, nobody understood precisely how disease was passed on, but it was understood that one could get a disease by contact—and it was not to be allowed! Even if each parish kept a special vehicle for diseased people, said the experts, this would soon be contaminated with all sorts of contagion. The answer was to reserve a vehicle, paint it a plain colour and pad it with hospital blankets which would become the patient's bedding. It could be easily cleansed.

## Buses and beyond

There was good money to be had from the 'buses'. Each week *The Times* listed the income of the London General Omnibus Company. In the week ending 23 January 1859, it was £10,191/11/1. There were also ways for dishonest conductors to make money on the side. This meant opportunities for several inventors too, who proffered clever counting systems during the year, designed to tally the passenger numbers so they could be compared with receipts.

It would take some time for them to be phased out altogether, but the writing was already on the wall for horse vehicles in 1859. A.L. Archambault was almost ready to run a steam-drawn tram on one of Philadelphia's railroads, using a two-ton locomotive, and steam carriages ran on the streets of New Jersey. They could reach quite a speed—an English steam carriage was clocked at 11 mph in Liverpool. In San Francisco, work began on 3 May on grading a line along Market Street, beginning at 3rd Street. On 28 December, a steam locomotive began moving on the line. A six-ton self-propelled steam fire engine carried nine passengers on the common road from Bristol to Philadelphia, covering 20 miles in two hours, and a 'steam elephant', an on-road traction engine weighing seven tons, had been tested on English roads.

It was too much for conservatives to bear. In an effort to regulate the speeds, the 'Red Flag Act' (the *Locomotive and Highways Act* of 1861) soon required British steam engines—threshers, steam rollers and the like, to move at no more than four mph in towns and two mph in the countryside, and to be attended by three people, one of whom had to walk 60 yards ahead with a red flag by day and a red lantern at night. The Act applied until 1896.

Jean Joseph Étienne Lenoir invented an effective internal combustion engine in 1859, and patented it in France and Britain in early 1860, but away from the cities, animal haulage was still the norm. In part, this may have been because Lenoir's engine ran on coal gas, which might have been a challenge to deliver to a moving vehicle.

Still, over the summer of 1859, 15,900 wagons, 12,000 head of cattle and 4000 people headed west along the Pacific wagon road to California. The road offered abundant grass, wood and water. On their arrival, a few of the settlers might have encountered another form of transport in California—camels. In late 1859, one Superintendent Beale at Fort Tejon in southern California reported to the Secretary of War about mules and camels and rated the camel superior in speed of travel and load carried. They were no harder to breed than cattle and, said Beale, he would rather have camels than three times the number of horses and mules. Unlike Australia, the US did not take to camels. But like Europe, the US *did* take to canals.

## CANALS, TUNNELS AND BRIDGES

By 1859, railways were out-competing European and US canals. Just 100 years after the Act of Parliament for the Duke of Bridgewater's first English canal was passed, the Netherton Canal opened near Dudley in 1859. It was the last British cargo canal, a finishing touch to remove a bottleneck on a congested stretch of water.

Gunboats had a single gun mounted in the bows, and massive armour (oak timbers and 5-inch iron plates) around the bows to protect the whole boat from whatever it was attacking—the boat just had to be pointed at the target and the gun fired. The vessels had one mast, a steam engine and a single screw, and appeared to be powerful weapons, but they were not perfect: two British gunboats 'were lately sunk in a conflict with the Chinese', said the

*New York Times*, probably meaning the four gunboats lost at the Taku forts in June 1859.

Gunboats were useful all the same, but with Britain still controlling the seas of the world, the French—who could not be sure of peace with Britain—preferred the inland route. A good canal could take the French fleet from the Atlantic to the Mediterranean and vice versa; a gunboat was taken from Marseilles by Toulouse to Bordeaux, 'nearly all the way by water', thus avoiding the British-owned territory, Gibraltar. *The Times* reported late in the year that a French government commission was looking into the cost of making the French waterways wide and deep enough to allow large vessels to pass through.

The emphasis everywhere had shifted to longer, wider, deeper canals. Work on the Suez Canal began on 25 April, and several routes were being explored for a canal from the Atlantic to the Pacific. In March, *Scientific American* reported that the '... mysterious Frenchman, Monsieur F. Belly, announces in the Paris journals that his organization of the Nicaragua Canal Company is completed'. This scheme, however, appears to have fallen at the first hurdle.

So too, did a scheme for a Spanish canal reported in the *London Illustrated News*. Sanctioned by the Queen of Spain on 25 March, this plan gave an Englishman, Charles Boyd, two years to prepare the way.

The route was from Bilbao—through the Cantabrian Mountains, the valley of the Ebro, past Estella and Saragossa—to the Mediterranean at the Bay of Alfaques in Catalonia. A large portion of the expenses for a canal 285 miles long, 340 feet wide and 30 feet deep—enough to carry vessels of the 'largest and most unprecedented dimensions'—was to be paid by the Spanish government.

The Suez Canal drew the most interest; in Britain because it offered a faster way to India, their most valued colony, and in the rest of the world, with so many canal announcements that never came to anything, because it was almost a reality. An old overland trade route had already linked Mediterranean and Red Sea ports until 1498. When Vasco da Gama found a sea route to India, it was virtually abandoned, but not entirely forgotten.

The advent of steamships made the old way feasible again, even before the canal. Passengers disembarked at Alexandria and travelled by a barge towed by a steam tug along the Mahmoudieh Canal and up the Nile to Cairo, where camels carried their baggage as the passengers rode in horse- or mule-drawn carriages to Port Suez to board another steamer. Before the Alexandria–Suez railway, the 250 mile distance was normally covered in about 88 hours. Once the railway opened in December 1858, the transfer switched to a faster and more comfortable mode of travel, but a canal would save the effort of unloading and reloading people, luggage and cargo.

All the same, not every canal was a success. In Indiana, said *Scientific American*, the receipts of the Wabash and Erie Canal had fallen from $193,000 in 1852, the first year of railroad competition, to just over $60,000 in 1859, not enough to cover working expenses. The answer, said the journal, was to use steam to move the boats faster. Steamboats were now active on the Erie Canal, and the old horse boats were 'threatened with annihilation'. The steamboats could propel themselves and haul several others. The steam tug

*Beemis* ran from Buffalo to Schenectady in five days and eight hours, towing three other boats its size. Steam canal boats offered one relief to operators: canal tunnels had no tow paths, so human power was needed to send the boats through tunnels—those on board needed to lie on top of the cabin or the cargo and use their feet on the tunnel walls or roof to 'walk' the boat through.

The earliest recorded use of 'tunnel' meaning a subterranean passage was in 1782, referring to a canal tunnel. By 1859, tunnels were associated with railways. One such was the Mont Cénis tunnel built for a rail link from France to Italy and begun in 1859. At the mouth of the tunnel, was a Belgian-made hydraulic compressor run by water power—compressing air to deliver power to the workface where it operated small portable drills to make blasting holes, as well as clear debris, and deliver fresh air to the tunnel head.

Cutting the eight-mile tunnel was planned to take six years, rather than the 30 years that older methods would need, but it really took  and was not completed until 1871.

Tunnels drew the attention of inventors, some brilliant, some eccentric, and a few of them completely unhinged. One French engineer of the third class offered a plan to build a tunnel running the whole length of the sandy waste of the Sahara, made by turning the sands of the desert into solid arched blocks.

The sands were to be dampened, formed into blocks, then fused by the heat of the sun with 'a huge Archimedean burning mirror'. Grateful travellers would be shielded from the heat of the day, and from sand storms, crossing the desert in safety. A side benefit would be the boost to Algerian trade.

Roads and road traffic usually went around water, or used small punts that could carry across several horses or even a coach. Trains were heavier, needing long and level spans that had to be stable under high loads, so the great names of engineering built the bridges. Robert Stephenson designed Montréal's Victoria Bridge and Isambard Kingdom Brunel completed the Saltash Bridge. These were serious structures, and the projectors of Saharan fused-sand tunnels and other mad schemes were not encouraged to apply.

The two great bridge builders, Brunel and Stephenson, left large fortunes when they both died during 1859. Both men's *greatest* bridges carried their first traffic in 1859, and still stand to this day. The consensus was that Stephenson was better than Brunel, as he was a practical mechanic as well as an educated civil engineer, and so understood all of the minutiae of the work. Brunel, on the other hand, is generally better remembered for a scheme of his that failed, the construction of the *Great Eastern*.

## STEAMING THE OCEANS

River steamers used timber as fuel but ocean-going ships needed coal. This fuel had to be carried aboard on the backs of sweating labourers which meant regular stops of 24 hours in coaling ports, and shipping companies made a feature of it. Schedules were geared to dropping mail, and the passengers were encouraged to go ashore and view 'the natives'. Quite inadvertently, a new style of tourism began to emerge, but people like Brunel began planning larger ships that could make long trips—like the run from England to Australia—without stopping for coal.

There were still fast sailing ships on some routes, but the statistics tell a story of steady decline in the number of these ships. In 1846, there were just two steamers plying the Atlantic; by 1859, there were 40, all much larger. In 1838, the first all-steam crossings took 18 days but by 1859, this was typically nine days.

Steamships in Britain (registered as steam tonnage on Lloyd's Register) totalled 104,460 tons in 1850, by 1855 the total reached 288,956 tons and at the end of 1858, 369,204 tons. In 1865, the steam tonnage on the register exceeded the sail tonnage for the first time. The month of May saw 22 steamships arrive at Boston, Quebec and New York, while 21 ships were expected to sail from Detroit for Europe in 1859, carrying staves, choice timber and flour. Future historian Hubert Howe Bancroft travelled by steamer and animal transport through Panama, from New York to San Francisco in 36 days in 1852.

As we have seen, Edward White took 23 days in 1859, which compared favourably with Bancroft's time. Each was far better than the best times of around three months for those rounding Cape Horn under sail, and steamers were getting faster.

Still, sail-powered schooners had reached San Francisco from Japan in just 23 days, and the original plan of Columbus, to approach Asia from the East, was alive once more. With a Pacific railroad, said *Scientific American*, a new trade and news route to the East was likely—but no new route would stay with sail for long. It was just too convenient to be able to head directly for your destination, ignoring the wind.

The age of sail had one last fling: the tea clippers of the China trade peaked in the 1860s and 1870s, the high point of the age of sail, even as it was foundering. Aside from being grown in China, tea was now cultivated in Java, Sumatra, Assam and Ceylon. There was a theory abroad that tea carried by sea was inferior to that carried by land, but Sir John Bowring, an old China hand, believed 'that tea loses none of its excellent sanitary qualities by being carried in ships; and in this opinion we think he is correct', said *Scientific American*.

The magnificent new steamship *Great Eastern* could have been a howling success: it certainly met the emerging need for fast mass transport, but the public could never see the need in time to save the ship, so it was a flop. Part of the *Great Eastern*'s problem was that it set too many challenges because it was so innovative.

It met no apparent need, it facilitated nothing, yet it was too amazing for anybody to ignore or forget it. The ship was too large to launch in the usual way, and had to be launched sideways. She got stuck on the slipway, there was a boiler explosion off Hastings and, later in the year, Brunel died. By January 1860, *Scientific American* reported that the *Great Eastern* had been sold off at half cost, and the acquiring company's shares were now selling at '50 cents on the dollar'.

The ship (originally to be called *Leviathan*) was judged perfect to carry and lay a new trans-Atlantic cable in 1866, but the demand for a 4000-passenger round-the-world steamer simply wasn't visible. In 1884, she was sold off to be used as a floating music-hall, and was finally broken up a few years later. Like the *Titanic*, the *Great Eastern* was rarely amazing or memorable in a nice way.

Around 1859, the safest bet for entrepreneurs was to launch a smaller steamer, and these began to show up in the strangest places. The Mississippi boats were booming in 1859, and on 6 April, Samuel Langhorne Clemens got his pilot's certificate after a two-year apprenticeship. In 1858, Samuel's brother Henry replaced him as crew on the *Pennsylvania,* only to die in a boiler explosion on board. Samuel—or as we know him better, Mark Twain—was very lucky to be alive.

There were steamers on the Red River, the Missouri River and the Colorado River, as well as the Mississippi, and sturdy spirits everywhere planned to haul smaller vessels in pieces to a point where they could be reassembled and used.

The launch of the *Great Eastern.*

There was even an iceboat being built at Prairie du Chien on the Upper Mississippi. This steam-driven sled, was 70 feet long with a beam of 12 feet, a driving wheel at the stern driven by two locomotive engines, a gripping rudder for steering and a powerful steam brake to slow it. The designer hoped to go at 40 mph, and to travel the 300 miles from Prairie du Chien to St Paul in a day, carrying 75 passengers over smooth ice and even over snow, up to three or four feet thick. If successful, wrote the reporter, it would 'introduce a new era in winter travelling in the north'.

The inventor, Norman Wiard, was a good engineer and no fool: he manufactured rifles and cannon for the Union Army during the US Civil War, but while the models impressed all who saw them, the real sled broke one of its runners while it was being prepared for trials. It never ran.

Small steamboats were highly successful everywhere. In March, Dr Livingstone's party was on the Zambezi river in Africa, where they 'dug some good coal' on its banks for their steamboat. No size was indicated, but one imagines a primitive *African Queen*. Steamboats on the Rhone were larger, up to 250 feet long, with a beam of 16 feet, and 500 horsepower engines. In Australia, riverboats were typically around 100 feet long.

## Australia's paddle steamers

The paddle steamer era began on Australia's Murray–Darling system in 1853, when two boats pushed up the Murray River, and showed a profit: one sold a cargo of flour, the other carried 4000 bales of wool downstream, doing many bullock drivers out of work.

By 1859, four river companies were trying to cash in on the trade, which lasted until the colonies of New South Wales and Victoria noticed their colonial produce was being carried off and sold through the South Australian capital of Adelaide, near the river mouth. Faced with a loss of profits, fees and taxes, the other colonies accelerated their railway construction programs. The Murray offered a special challenge. The riverbanks were lined with river redgums that gave marvellous fuel for steaming, but which dropped branches and whole iron-hard trees into the stream, where they sank and lurked as snags. Other trees stood on the banks, alive or dead, jutting out over the river, waiting to sweep away the superstructure of any boat that came too close.

Americans who were plying the Murray in Mississippi-style structures learned the hard way that not everything transfers from one place to another. The surviving boats were lower side-wheelers which hauled barges, often several of them, as large as the boat itself.

Even fully laden, few of the vessels or barges drew more than four feet, but the cargoes changed country life forever. The steamers carried furniture, pianos and machinery up the river—and at least once, an elephant. The more common cargoes inland included riveted iron boxes, filled with breakables or valuables—china, fabric, pottery and tea. The iron boxes served to protect the cargo from water damage and were sealed to make pilfering difficult and tampering obvious. Downstream, the loads included bales of wool, compressed in steam presses and scientifically stacked on the decks or in the barges, getting the wool to market and making profits for the graziers. Along the river, poor families scratched a living by cutting dead trees and selling the wood as fuel.

Australia's Darling River was even harder. While the flow in the Murray was predictable, fed by the thaw of snow in the high country each spring, the Darling drained country with no real annual seasons, where occasional monsoonal rain depressions from the north could deposit water in the channels and plains that fed the river—and the plains went on forever. Huge amounts of water gathered, far upstream of the navigable river.

Water might appear in the Darling, pouring down from a wet north a month or more after the rainfall had ceased, filling the riverbed to its banks, and spilling across the plains and out into parched countryside that had, usually, seen no rain for years. When the waters ran like that, paddlesteamers could get 2000 miles from the sea and, according to tradition, up to 50 miles from the river, which was well and good if the boat got back into the river channel before the water fell away again. If it missed finding the river, the landing was gentle and the boat could escape in the end when the next flood came—more than you could say for many less planned marine landings and strandings.

## SHIPWRECK!

In 1859, the Plimsoll Line—which enshrined in law a minimum, safe load line—was still many years away, and greedy ship owners fought any attempt to stop 'coffin ships' sailing with too little freeboard. Coffin ships got their name because they were overinsured and worth more sunk than afloat. Generally, ships and cargoes were well insured, though normal life insurance policies issued in Britain were rendered void if the policy holder travelled by ship without special permission. The life insurance companies knew the odds.

Until the age of steam, it was easy for a ship to be driven ashore in a storm, for a small navigation error to take a ship to where wind and waves forced it ashore. The Caribbean and Cape Horn trapped many vessels, and South Australia's coast was like a scoop, slanting to the south east and gathering up sailing ships running before the westerly 'roaring forties'. Even steamers were not immune to coastal traps and currents.

*Admella* was a steamer, built in Glasgow in 1857 to serve the Australian inter-colonial traffic. The ship was 360 tons, had twin 100 horsepower engines, and started on the Adelaide–Melbourne run in June 1858. She left Adelaide on 5 August 1859, and picked up a fireman and a few more passengers, giving her 82 passengers (including 19 women and 15 children) and a captain and crew numbering 31.

The ship carried 93 tons of copper and seven horses, four of them racehorses set to run in an inter-colonial championship race in Melbourne on 1 October. The horses were in boxes on the deck, and when the ship went through the notorious Backstairs Passage between Kangaroo Island and the mainland, the trouble started.

Lumping seas are the order of the day there, and the rolling was enough to throw one of the horses on its back inside its box. The ship was brought around to head into the swell while the horse was calmed and helped into a standing position, but it seems the crew became disoriented. When they got under way again, they drove their vessel onto Carpenter's Rock. The ship broke in three, with the stern section being held in place by the dead weight of 60 tons of copper as the rest of vessel was carried away.

Several men drowned while trying to swim to shore; two other ships sailed by without even noticing the wreck. When the weather improved two men lashed together a raft and reached the shore. Eighteen men and one woman clung to the weighted-down stern for seven days and nights before they were finally rescued.

The wreck of the *Admella* was a colonial scandal and cause for concern, but the loss of the *Royal Charter* in late 1859 was a far greater concern in Britain. Charles Dickens visited the site of the wreck and wrote about it in *The Uncommercial Traveller,* while every British man and his dog held an opinion on the cause. *Royal Charter* was a famous iron ship which had made the journey from Liverpool to Melbourne in 59 days.

The Reverend Captain William Scoresby FRS had travelled to and from Australia in the ship in 1856, studying how the compasses behaved, because compass adjustment was still an inexact art at best. Scoresby, a famous whaling captain who later became a parson and scientist, wanted to study the interplay of terrestrial magnetism and compass deviation on board an iron-hulled ship, but achieved little before he died in 1857. The meagre results of his researches were published posthumously in 1859.

Aside from the link to Scoresby, the ship's owners also made her famous, advertising her speed. *Royal Charter* was 2719 tons, and had a 200 hp auxiliary engine, so the ship was nearly eight times the weight of *Admella,* but she had the same engine power. *Royal Charter* also had sails, and they became part of the problem.

On 26 August, the ship left Hobson's Bay in Victoria bound for Liverpool with half a million pounds (around $100 million today) worth of gold and 400 passengers. Reaching Ireland in late October, she anchored off the Cove of Cork, and some of the passengers sent letters and telegrams, by the *Petrel* pilot boat, to their family and friends. Then, as the weather closed in, the captain pushed on for Liverpool.

The wild storm on 25 October wrecked more than 200 vessels, but one ship grabbed people's attention. Off the Skerries, the *Royal Charter* signalled for a pilot, but no pilot could put out. The ship anchored, but both cables parted in quick succession. Wind pressure on the masts and rigging probably added to the strain, but the skipper may have avoided getting rid of the masts, fearing that the stern screw might be fouled by some of the lines that would still be attached—or perhaps he thought the masts would not go over the side cleanly, in which case they could have tilted the ship or swung it around.

The ship struck and the masts were then cut away, but it was too late. Guns of distress were fired, blue lights were sent up, and a line was put ashore. The captain sent 16 crew members to work the line, but before anybody could be landed, the ship broke up. Only 39 of the 498 passengers and crew survived. They had travelled for two months, and were within two days of home, but they never made it, and Britain was aghast. A writer styling himself 'Amicus' wrote to *The Times* on 6 December, arguing that iron ships were made of poor quality iron called 'boat iron'. The writer wondered why a ship, a mere 50 yards offshore with a hawser in place, saw so many deaths. The *Great Britain* had stayed aground for a whole winter without breaking up, and, said Amicus, other examples showed that a well built iron ship was safe, but *Royal Charter* was made of materials as wrong as a Yankee trader's wooden nutmegs, a grocer's sanded sugar or a petty swindler's sewing cotton that is shorter than advertised.

Three days before the Amicus letter, *The Times* reported that the *Royal Charter*'s plates had been tested before departure and been found to be 'above standard'. The view now is that the captain of an ordinary sailing ship might have dropped anchor, but the proud skipper of a famous fast ship felt impelled to rely on a drastically underpowered engine which could only drive the ship at eight knots per hour ('knots per hour' was the common usage in 1859) in dead water, according to *The Times*. It was a learning experience, but a harsh one.

## Defences against sinking

Two ingenious New Yorkers, H. Hallock and Isaac Smith announced in *Scientific American* that they had designed a stateroom, self-contained and sheltered below a deck that might open, a room able to be sealed and equipped with food and water for those within. In an accident, the stateroom would be detached and allowed to float free of the ship. It had a pump to keep waters at bay and lamps that could be lit at night. The idea sank without trace.

Other inventors were determined that warships, which by definition carried explosives, would be safer than they had been to date. In France, *La Gloire*, described as the world's first iron-clad ship, was launched in 1859, though Korea had iron-clad ships centuries before, in the late 1500s. A ship of the line in the early

19th century used 3500 oaks, the product of 900 acres, timber which needed to be seasoned for up to 25 years. But saving trees was not *La Gloire*'s inspiration, because she was timber beneath the plating. The ship was a response to the burning of the wooden Turkish fleet at Sinope in 1853 by the Russian navy's explosive shells.

The launching of *La Gloire* signalled the end of the wooden warship, because where one nation led, others had to follow. In 1860, the Royal Navy's HMS *Warrior* had guns mounted on a single deck, running 380 feet, displacing 9000 tons.

The last wooden British ship-of-the-line was a three-decker launched in 1859, but by the end of the century, timber ships had been replaced by iron and steel dreadnoughts weighing 20,000 tons.

Charles Atherton, chief engineer at the Woolwich Dockyard wrote to *The Times* in January 1859 to share his idea for making vessels with an interior of light material up to the waterline. Gunboats, floating batteries and mortars could benefit, he said. The solids might consist of 'cork shavings, light wood sawdust, rush stems, cotton waste, flocks, hemp, and other lightweight material, which, by the aid of a solution of gutta percha or other chemical process, would form a solidifying mass, so tough that it could not be knocked to pieces by shot, and so light that it would only be one half the specific gravity of water, and therefore, unsinkable, however perforated by shot ...'

Atherton had previously offered a similar idea for treasure ships, so they would float, allowing recovery of the riches. It took until the end of the year for *Scientific American* to mention the idea, when a writer responded saying that cork would not suit because heated shot could set fire to it, but a suitable material ought to be able to be found. Half a century later, most lifeboats were fitted with sealed cork-filled compartments and self-draining seacocks to keep them afloat under the worst of conditions.

Even if water transport in tea clippers was supposed to harm tea, the French had other ideas about moving wine. Late in the year, *The Times* reported that the levels of the upper Seine, Yonne, Loire and Saône had increased and more water was now flowing in them. This improved navigation meant more boats, and so more wine was being delivered at Bercy. Water carriage, said the reporter, was preferred to railway movement, which was believed to interfere with the flavour of the wine. The Parc de Bercy was a major wine depot, linked to the wine regions by both rail and river, but outside of Paris, and free of irksome city taxes because it lay just beyond the fortifications of Thiers, named for the same minister who had once dismissed railways as 'toys for the curious'.

## TAKING TO THE AIR

Rockets and flares were once toys for the curious, later they were used as generic signals of distress, but in 1859 they were being used for more complex signals. Women's names were beginning to appear on patents, though usually on inventions related to the household, so Martha J. Coston broke new ground when she applied for a patent on a maritime signalling system in 1859. Though she sought the patent as executrix to her late husband, he had, in fact, left only the bare bones of the idea, and she needed to do much of the work herself. Designed to let ships communicate at night, the signalling system sold in a number of countries, and also to the US Navy in the Civil War, where a typical signal was 'Blockade runner going out: One rocket followed by Coston's No. 0'.

Another airborne item was the balloon. Now that coal gas was more common, more people could fill and launch balloons, and were finding new uses for them, though some uses were far from new. Balloons had been used in the 1790s to spy on enemy action, but in 1859, the French used balloons in their war against Austria. Two years later, Union forces in the US used balloons in the US Civil War.

In the US, balloonists hoped to travel far greater distances, even to cross the Atlantic. In July, John La Mountain and three other men set a world record—they drifted 809 miles in a balloon from St Louis to New York; the record lasted until 1910. Scientist and inventor, Thaddeus Lowe's projected trans-Atlantic balloon was being assembled in New York. It was reported to have a diameter of 130 feet, a height from top to basket of 350 feet, a weight of three-and-a half tons, a lifting power of 22½ tons. It was said to hold 725,000 cubic feet of coal gas, which would have cost something like $2 per 1000 cubic feet.

According to *Scientific American*, 17 sewing machines were used to make the envelope, with three layers of cloth and triple seams at the top of the bag. The cloth was varnished to make it gas proof, and attached to a rattan basket, 20 feet wide and four feet deep, warmed by a lime stove that would provide heat without flame.

Lowe hoped to cross the ocean in 48 to 64 hours, landing in France or England, rather than Spain, which lay due east of the

Thaddeus Lowe's trans-Atlantic balloon.

ascent point. For safety, the balloon had a metal lifeboat under the basket, with an Ericsson engine driving 'fans' that should, he said, give some degree of steering power. A 10-inch gas main, supplied by the Manhattan Gas Company, was used to fill the balloon.

While being inflated, the balloon was housed inside a high board fence; to help defray expenses, there was a 25-cent charge for admission. On show were the lime stove, a grappling iron for stopping the balloon, and one of the copper vessels that would carry compressed gas to replenish the balloon's supply.

In the end, the flight was cancelled. The launch needed 60 hours of fair weather, which simply had not eventuated. Lowe still hoped he might reach Britain's shores before Brunel's *Great Eastern* reached America, but it was never to be. Though he may have thought himself unfortunate, he may well have been lucky, considering John La Mountain's second flight.

La Mountain started from Watertown, New York, on 22 September with John Haddock, a newspaperman. They went almost straight up to 3000 feet, then drifted northeast at 25 miles an hour, keeping this rate up even when they were three and a half miles up. Soon they were above the scudding clouds, with no idea how fast they were travelling, as they drifted along with the clouds, keeping a more or less constant position with them.

They moved up and down several times, uncertain as to where they were, and on one occasion, Haddock grabbed onto a treetop, which they recognised as spruce. This suggested they were almost certainly in Canada, far north of where they had thought they were. It was night, so they made fast to the spruce, to await daylight. When day came, they were soaked by rain; as was the balloon, which La Mountain believed had gained 100 pounds in weight. To lighten the load, ballast, a blanket, porter bottles, a coat and an anchor rope went over the side.

Away went the balloon, only to settle later, even further into the woods. They struggled through the forest, rafted down a stream and, eventually, were lucky to meet a party of lumberjacks who helped them. Without that fortunate encounter, they would probably have died.

Either that knowledge or the looming Civil War put a bit of a dampener on further balloon travel, though balloons were certainly used as tethered observation points during the war. In the panic that followed the first Battle of Bull Run, Lowe used a balloon—he went aloft to see if the Confederate forces were moving on Washington DC, but thereafter, balloons were a minor military item. The future of flight, or one part of it, however was sealed. Hugo Junkers, future aircraft designer, was born on 3 February, and Lenoir's internal combustion engine was completed by the end of the year.

Aircraft would not fly until small and effective engines were available—and that would require experience and understanding. Experience would come with time, understanding required hard work and a willingness to accept that sometimes, just sometimes, science is counter-intuitive. Evolution *does* happen, rockets *do* work in a vacuum, heavier-than-air machines *can* be made to fly, and perpetual motion would *never* work. The real challenge for some lay in understanding that energy is never free.

Some still do not understand this, even today. This is why we still hear of motor vehicles using water for fuel, free energy, and other curious or spurious claims.

# 4: Understanding energy and power

*If my views be correct, a fall of 817 feet will of course generate one degree of heat; and the temperature of the river Niagara will be raised about one fifth of a degree by its fall of 160 feet.*

James Joule, *Philosophical Magazine*, 1845.

On 1 January, *Scientific American* addressed the question of units of heat. There were, the writer said, several possibilities, but it was now clear that heat was an energy not a substance. James Joule had shown in 1843 that heat had a mechanical equivalent, which was scientist-speak for 'weighing hot bodies will never find them any heavier than cold ones because heat has no mass'. Before Joule, heat was thought of as some sort of fluid with mass. Once that misperception was clarified, once scientists and engineers understood energy, science surged forward.

All our modern mastery of energy was within reach in 1859, though it was pinned down and dissected out over the following three decades. In 1859, William J.M. Rankine worked out his Rankine cycle which was important to obtaining efficient power from steam engines. It was the year James Clerk Maxwell determined his distribution law of molecular velocities; the year Gustav Kirchhoff related black body radiation to temperature and frequency.

Maxwell's work related the temperature of a gas to the average velocity of its particles, but also allowed that the atoms and molecules would have a predictable statistical distribution of speeds. By extension, this explained all sorts of things, even the way that wet clothes get cold on a windy day. Kirchhoff's work offered fewer commonplace corollaries, but it set a theoretical basis that would flower into quantum physics, half a century on. Physics was getting serious.

All the same, when writers in 1859 discussed food as 'the fuel of animals', they only mentioned heat needs, saying people in cold climates needed to eat more. The notion of food providing the energy to perform work was a step too far. The knowledge existed, but it still had to spread.

It was getting there. In November, *Scientific American* said perpetual motion would only work in a very limited sense. There were seven forces capable of moving matter: Heat, Gravitation, Muscular Power, Magnetism, Electricity, Chemical Affinity and Capillary Action.

The writer allowed that some of these might be shown to be the same, and that light 'or other forces' might need to be added to the list, but only three had any prospect of delivering power perpetually: gravitation, heat and muscles.

In this context, a water wheel showed perpetual motion, so long as water was available. Tidal energy, based on gravitation, also delivered perpetual motion, as the writer saw it. Equally perpetual was the sniping between journals on the topic. Six weeks earlier, *Scientific American* had denounced the *New York Tribune* for endorsing a scheme for a self-powered electrical generator: 'If electricity could be applied to produce such effects, so could wind, steam, water and animal power . . .'

In July, the journal reported that a Professor Salomons was using 'bisulphuret of carbon' (carbon disulfide) to replace the water in the boiler of a steam engine to make it more efficient. Salomons was furious because he believed that his invention was being used in Europe without any credit or payment to him. He claimed to have treated the substance to remove its smell, but as it is fatal if inhaled, swallowed or absorbed through the skin, one wonders how this lack of smell was assessed.

## KING COAL

A numerate Lancashire coal miner calculated that the 68 million tons of coal raised in Britain in 1859 would make a hole six feet deep, 12 feet wide and 5128 miles, 1090 yards in length, or a solid globe with a diameter of 1549 feet. Coal was a growth industry: in Britain, 10 million tons of coal were mined in 1810, 100 million tons in 1865, 200 million in 1875. But it was also a dangerous industry. Around 1000 miners were killed in accidents each year, though fatal accidents were unusual—some 20 per cent of miners suffered an injury each year, but most survived.

When the Main Colliery near Bryncoch, about two miles from Neath in Wales, flooded, it killed '25 men and boys and several valuable horses', said *The Times* on 9 April. The miners had been digging a second exit from the mine, as recommended by the Government Inspector. They cut into the nearby Fire Engine Pit, which they thought had been thoroughly drained, but this was not the case. Water poured in, the alarm was sounded, and 55 men and boys, along with two horses, were saved. The horses saved themselves by leaping into the tub as it was lowered, the men said.

Nine days later, pumps were put to work to lower the water, but it took six months, until October, before the pit was drained, allowing workers to recover the bodies, which were described as 'now greatly decomposed and only identifiable by their clothing'. At least, said people, no women or children were killed: the *Coal Mines Act* of 1842 had barred women, and children under 10, from working underground. Men kept going down into the mines, so long as the miners' health lasted, because the money was good. The money was good because coal could be used to warm the homes of the rich, and to make steam, and coal gas, and coal oil.

Patrick O'Brian's fictional naval hero, Jack Aubrey, had a real-life model, Thomas Cochrane, 10th Earl of Dundonald, who rose to be an admiral in the Royal Navy. When Cochrane's autobiography was being ghost written for him in 1859, some of 'his' words would have been penned under gaslight; the same gaslight which his father, the 9th Earl, had discovered while attempting to distil coal tar. The inventive older earl hoped the tar would help preserve ships' timbers and rigging.

The experimenter saw that the vapours were flammable, collected some in a gun barrel and lit the gas. This produced a satisfactory (to him) flare of light, visible on the other side of the Firth of Forth, where nervous people were probably rather less satisfied. Cochrane senior was hoping to make money just from his tar, and seems to have missed the possibilities of coal gas, and unlike some, he paid a high price for his single mindedness. Dogged determination and pigheadedness can only be distinguished in hindsight by looking at the rewards reaped.

William Murdoch invented coal gas lighting some years later in 1804, so gas was mature technology by 1859. There were 1000 gasworks in Britain to supply gas for lighting, and 297 companies making gas in the US. Gaslights were fitted during the year in the offices of *Scientific American* in New York and the new Houses of Parliament in London. The *American Gas Light Journal* began publication in New York in 1859.

Gasworks at the St Denis Hotel, New York.

Coal gas in London was 4s 6d per 1000 cubic feet, while New York gas cost $2.60, the equivalent of 10s 6d. Cannel coal from Britain, the main source of gas was the same price in each city, even though it had to be hauled across the Atlantic to New York. Gas mains were kept short, so there was no call for large-scale gas generation in the year, but *Scientific American* advertised village-sized gasworks for sale.

Australian towns also used British coal, though an unconfirmed legend has it that the Victorian town of Kyneton obtained its gas from eucalyptus leaves for a while.

The gas makers charged what the market would bear. So the St Denis Hotel on the corner of Broadway and Eleventh Street in New York became the proud owner of its own gasworks, producing gas for illumination at less than $1 per 1000 cubic feet. The coke by-product was used to fuel the process, with any excess going to the kitchens. In all, 500 lights were supplied, and it took half of one man's working time to operate, maintain and service the system.

Pilsen, in today's Czech Republic, had gas lamps in its main streets and square from 1858, gas pipes were laid in Istanbul in 1859; streets 'heretofore almost impassable after dark' could be safely walked.

Athens had also been lit, and Rome acquired gas technology from England, once her furthest and most insignificant colony. Even Honolulu and Sydney had gaslight. When a Miss Johnson published her *Geography with useful facts for the junior classes in schools* in 1859, one of the facts that children in Australia were expected to know was:

> Q: How is Sydney lighted?
>
> A: By coal gas, which is made at the works of a private company.

During the year, *Scientific American* reported, in some wonder, that a Dundee church was being heated by gas in stoves. Dundonald would have been pleased that coal tar was finally coming into its own. The Royal Navy had started using tar from gasworks on their ships, though they (incorrectly) credited Sir Humphry Davy with the idea. And in 1856, William Perkin used coal tar to start a whole new industry—making the dye, mauve.

## OIL

Whales were hunted to provide whalebone and whale oil but, as Edward Forbes noted in his 1859 study of the seas, '... whale's blubber cannot put up with incessant persecution any more than human flesh'. This persecution was, however, still pursued with the greatest possible efficiency. In August, *Scientific American* reported that a harpoon had been invented which made a whale kill itself: each pull drove the harpoon deeper into its flesh. Another inventor offered an electrical harpoon that shocked the whale to death.

Whale oil was an essential item in the 19th century. It lubricated the machinery of industry and it provided the light by which scholars learned, invented and wrote paeans of praise to the accomplishments of humankind in taming Nature and directing its wonders to their ordained purpose of serving human needs. Forbes was one of the very few who could see where exploitation would lead.

He noted that the whale fisheries between Spitzbergen and Greenland had been reported as abandoned some 20 years earlier. From about 1850, the Japanese learned the problems of global trading when western (mainly US) whalers began taking whales in Japanese waters. In May, the *New York Times* reported that the spring fleet of whalers was returning to Honolulu with a far from satisfactory catch.

Still, *Scientific American* had good news for whalers that was bad for whales, the discovery of a great laguna in Lower California.

About 100 miles long, ranging from 28°4′ north down to 26°40′ north, its waters were teeming with whales and seals, but, 'It is not likely that they will prove so abundant after the approaching whaling fleet has killed off the cow whales, or driven them from their ancient haunts', said the reporter. The area was probably Laguna San Ignacio, which ranges from about 27° north down to about 26°40′ north, and now a centre for whale watching. The error in the latitude was probably deliberate, intended to trap copyists.

At least one sperm whale was fighting back, attacking whaling boats and ships. Whalers called him 'Mocha Dick', after Chile's Mocha Islands, where he made his first attacks around 1810. The inspiration for Herman Melville's *Moby Dick*, he was grey rather than white, though he had a large white scar on his head. A Swedish whaler killed Mocha Dick off the Brazilian banks in August 1859. Measuring 110 feet in length, he weighed more than a ton for each foot, but the whalers' nemesis was captured without much of a struggle. According to the whaler, the animal was blind in one eye, almost toothless and dying of old age. A bit like the whale stocks, you might say.

The Atlantic whale fisheries were operated by steamships, but the haul from New England ports of the US to the Pacific was still in the hands of fast-running American sailing ships. Needing no coal, they could stay out until food or water ran low, or their holds were filled with brimming barrels. In 1834, the US owned 400 of the world's 700 whaling vessels; by 1859, it was 661 out of 900. The value of the catch had risen from $5.5 million to $12.3 million. Sperm oil harvested was up from 95,000 barrels to 193,300 barrels, whale oil had risen from 146,500 barrels to 153,800 barrels, and whalebone from 1.175 million pounds to 1.538 million pounds.

The first oil well was drilled in 1859, just as whale oil production started falling, so people often suggest a spurious causal effect, but the truth is more complex. There was 'crude oil' long before there were oil wells, though it only came indirectly from the ground. Coal could be turned to oil as easily as it was made into gas.

Even so, it was coal gas that did the real harm to whaling. Gaslight was cheaper than the light of whale oil lamps, and in the end, most British whalers were converted for coal carrying. Whale oil, especially sperm oil, was, however, still highly regarded for lubrication in machinery, and was largely reserved for that purpose. Whale oil lamps smelled less than those burning other animal fats, but kerosene made from coal oil was much cheaper. *Scientific American* explained, in a July article, that a decade back, almost all lighting oil was whale oil, but now, just nine years after the art of making coal oil was mastered, five years after it began in the US, 'more oils are made from coals in one week, in our country, than ever was obtained by our whale-fishers in the best year's fishing they ever enjoyed'.

## The rise of coal oil

James Young of Manchester, England gained a patent for making oil from coal in 1850. The best coals for oil making were cannel coal, parrot coal and gas coal, the same coals normally used to make gas. By heating the coal at low red heat (below the heat required for making gas), the makers could capture most of the volatile components as liquid. To purify the oil, Young treated the raw liquid with dilute sulfuric acid which he neutralised with caustic soda. Next, the oil was distilled with water and chilled to 40°F to extract 'paraffine' (wax) which could be filtered out by passing the fluid through wool or other cloth.

In 1859, the Admiral commanding the Royal Navy's North American and West Indies station was the 10th Earl of Dundonald, Thomas Cochrane, who shared his late father's interest in oily, coaly things. Cochrane's role took him into the warm waters around British possessions in the Caribbean and also to Canada.

This is how he had obtained the rights to the pitch lakes on the Caribbean island of Trinidad and then met Abraham Gesner, a native of Nova Scotia who had invented a process of extracting kerosene from bituminous coal.

Cochrane thought the Trinidad bitumen might make a good road surface but people said horses would slip on it, so that was dropped. He also considered it as a building material, as insulation for cables, and as manure. In this last, he was following tradition: in 1816, his father had suggested shipping British peat to the Caribbean as manure for sugarcane, now Cochrane wanted to take bitumen the other way, but in the end, he proposed using it as a fuel source.

Trinidad was a source too distant, but the two men got on well, and Gesner continued developing refining methods he had invented for the bitumen. He applied them to coal, but James Young then successfully sued Gesner for breaching his patent. Then Edwin Drake and others found oil, first in Pennsylvania and further afield. Still, Gesner's techniques were well known, and formed the basis of the petroleum industry.

Before the development of 'rock oil', coal oil was highly profitable. The Lucesco Oil Works in Pittsburgh cost $150,000, employed 150 hands, and in 1859 turned out 5000 gallons of crude oil in each 24 hours, getting 40 gallons of liquid from a ton of Pennsylvania cannel coal. This stock yielded 30 gallons of crude oil, or 24 gallons of finished oils for burning and lubrication. A gallon of oil, extracted from a bushel of coal costing five cents, was selling for 35 cents or more.

The oil from wells was called 'Seneca oil'—the Seneca people had traditionally collected the oil from seepage points. The Pennsylvania Rock Oil Company bought rights to an oil spring for $5000 and leased it to Edwin Drake in exchange for a royalty of 12½ cents a gallon. By 1859, drilling technology was ready for the new demand: one artesian well at Louisville Kentucky had been driven 2086 feet below the surface. Drake had a small hole drilled down some 70 feet at the oil 'spring', raising the yield from 400 gallons of oil a day to 1600, according to the *Erie Gazette*.

Up until then, reported *Scientific American*, the oil was known to surface at Amiano in Italy, Birmah, on West Indian islands, and in Virginia, Kentucky, New York and Pennsylvania. Most uses were medical: it was used for chilblains, rheumatism and paralysis, given both internally and externally.

The oil from Pennsylvania sold for 60 cents a gallon in New York. Pumping up a gallon of oil cost about a cent, and barrels and transportation were estimated at 9 cents a gallon, leaving a good margin for profit. Even a New Lisbon, Pennsylvania 'coal-oil spring', yielding five to eight gallons a day, would offer enough profit to keep a family in comfort. With Gesner's work available to others, generating oil fractions to burn in lamps was an easy matter. Coal, they reasoned, was not oil; so James Young's patent did not apply.

## THE SCIENCE OF LIGHT

The lighting of New York was inadequate, complained the *New York Times* on 1 January. It favoured pickpockets and thieves because the lamps were too dim, too far apart and, sometimes, blocked. In Brooklyn, the street lighting was no sham, it said, but across the East River, one plunged into 'Egyptian darkness'.

Comparing the effectiveness of different forms of lighting was less than easy, but the scientists went to work. First came Robert Bunsen, best known for the Bunsen burner that he did not invent, with his Bunsen grease spot photometer. This simple device used greasy paper to compare two light sources, but the real need was for reference sources, standard lights that could be used in different places and at different times.

The official British reference point for light intensity was based on a standard candle made of spermaceti oil with a small amount of beeswax added to control brittleness, and a melting point of 112°–115°F. The French standard used a Carcel lamp burning colza (rapeseed) oil at 42 grams per hour, with one Carcel unit equivalent to about 9.5 British candles. Neither was particularly satisfactory as a unit, but they allowed measurement of a sort. A comparison in *Scientific American* of the cost for an equivalent supply of light from different sources showed that kerosene costing $4.10 gave the same light as \$4.85 of camphene, \$12.00 of whale oil, \$17.70 of lard oil, \$26.47 of sperm oil and \$29.00 of 'burning fluid'—a patented mix of alcohol and turpentine.

Dr Andrew Wynter's essay on Price's Patent Candle Company at Battersea described a progressive factory with a school room where the company trained selected brighter lads 'for those departments requiring particular attention'. In his methodical way, Dr Wynter, a physician and author, explained the many oils and fats that could be used for light—'shea butter', an oil extracted from the fruits of the African shea tree, petroleum of Ava, 'the beautiful insect wax of China' and even tiger fat. Petroleum, he said:

> ... comes from the kingdom of Burmah, where it wells up from the ground like naphtha, which it resembles. It is a dark orange, and can be distilled to give sherwoodole, a detergent used to clean gloves and remove grease stains. A lamp oil, Belmontine oil, is next: this burns clear and bright, then there are two qualities of lubricating oil and a wax called Belmontine, used to make candles similar to those made of paraffine distilled from Irish peat.

The price for candles in New York varied rather dramatically. At the end of the year, sperm candles were 35–40 cents a pound, patent sperm candles 50 cents, paraffin wax candles 50 cents, and stearic candles were 27–28 cents. On the strength of that, the candle business might have looked good, but electric light was coming. Each evening in July, Professor Moses Farmer lit one room of his Salem, Massachusetts house with lamps that used small pieces of platinum-iridium wire which glowed dimly as the current from primary batteries passed through them, though these were probably not the lead-acid storage batteries that Gaston Planté developed during 1859, and which we still use today.

A steam-powered generator lit, for a short while, an electric arc at a Foreland (England) lighthouse on 8 December 1858. In March 1859, a second arc light was added, along with two more at Dungeness and tests of lighthouse arc lights continued in Britain and France for two decades.

In Paris, an arc light was mounted on the back of a wagon and drawn through city streets at night in the summer of 1859, together with a 'magneto-electric machine' (a steam-powered generator), though it was noted that currents of electricity could be generated by a stationary steam engine and 'carried to the point of use by metallic conductors'. The concept of electrical lighting was all there, even if the execution was delayed.

Lighting was important on stage, where players were, by now, literally hogging the limelight—otherwise called the calcium or Drummond light. The limelight effect was discovered by Goldsworthy Gurney of Cornwall, who mentioned it in a chemical work some years before Thomas Drummond used it to make signal lights for his trigonometrical survey work.

Gurney was happy to refer to it as the 'Drummond light' in its developed theatrical form, though he was proud to claim credit for his 1820s steam road carriages, despite their failure because before the middle of the century, the engines might work to pull loads on rails, but they were too slow and heavy to work on roads. The faster engines which were becoming the norm by mid-century required better lubrication.

By 1859, the lack of suitable lubricants was hampering progress at sea and on land. Steam engines, by that time, turned at 60 rpm, enough to demand oil lubrication, rather than the quick dab of grease or tallow that had been the practice to date. Rapeseed oil worked, but linseed oil and most of the other vegetable oils were 'drying oils' that would eventually leave a hard coating on surfaces, gumming up the works and causing problems. Coal oil was too volatile, and 'earth oil' or 'rock oil' was still too rare.

Rock oil, said a November report in *Scientific American*, was regarded by machinists as the best lubricant available, but opinions were mixed. On the other hand, a study on the Michigan Railroad had been reported in the *Railroad Register* in September. Engineers had compared whale and 'metallic' oils, and favoured whale oil. Running a single train 103 days, one half of the journals were lubricated with whale oil, consuming 28½ gallons, costing 60 cents per gallon; the other half used metallic oil, consuming 27 gallons costing $1.34 per gallon.

At least one thinker considered the use of oil in a new way. The possibility was mooted of using 'crude oil'—that is, unrefined oil from coal—as steamship fuel. The oil equivalent of coal would weigh half as much, which would mean more freight capacity or a greater range and there would be a significant saving in labour.

Two weeks later, a correspondent to *Scientific American* suggested that resin oil might be even better because it stayed liquid at low temperatures, while crude coal oil needed to be cut out with a shovel, or steamed out. Nobody suggested using 'rock oil' for fuel. That sort of thinking would not emerge until petroleum fractions were more common.

## THE MACHINE AGE

Machines were in use everywhere: there were even clockwork-powered fly and mosquito traps. Items once made by hand—shoes, for one—were now being turned out of machine-filled factories. Patents were granted for machines that did perfect mortice-and-tenon work in furniture, and machines were taking over the toil of harvesting. The superiority of steam over human or horsepower for threshing, grinding and cutting fodder and roots was declared by *Scientific American* suggesting farmers with 100 acres or more would soon turn to steam, though only if ' ... upon careful experiment, it can clearly be demonstrated that it is more economical ...'

By year's end, an assumed greater public understanding of energy is visible in a *Scientific American* report, offered without elaboration. Readers were told that Nasmyth's steam pile driver, operating at 60 to 70 blows a minute, sometimes caused the head of a pile to burst into flames. To make sense of this, readers needed to recognise that mechanical energy had been converted to heat—and the reporter clearly believed readers would be able to work that much out.

The advertisement pages of *Scientific American* tell a wider story. Items offered for sale in those pages at the start of the year included a stave-cutting machine, steam gauges, cotton machinery, oil 'that will not gum', steam engines, books, journals, pipe, a brick machine, lathes, planers, a water mill site, scales, flues, felt for steam pipes, grain mills and coal oil retorts. Items on offer at the end of the year

A well equipped and active machine shop.

included hosiery knitting machines, planing machines, lathes, steam pumps, wind turbines, wood-bending machines, circular saws, hay presses, gas works, grist mills, clocks for churches, machine and burning oil, shingle machines, machine belting and boiler flues. Books included a complete guide to distillation, a life of George Stephenson, Liebig's *Modern Agriculture* and a manual on topographical drawing.

There were also assorted inventions for making sugar, shaping projectiles for rifled ordnance, forming hat bodies, as well as railroad spikes and sewing machines. The sewing machines were perhaps the key innovation of the decade. That ornate monstrosity, the crinoline, would surely have been too expensive for most women without the saving effected by the sewing machine, which was equally useful for making the giant bags of coal-gas-filled balloons. Invented in the US in the 1840s, the machines were slow starters in Britain, but by 1865, chain-stitch and lock-stitch machines were common there. Rather than displacing seamstresses, the machines made production more efficient and allowed the same effort and time to produce clothes with more ruffles and decorations.

William Newton Wilson introduced the first domestic British sewing machines to the market in 1858, though Elias Howe had been selling US models in London as early as 1847 and US machines continued to compete, joined by German machines in the 1870s. By 1859, in Britain and the US, sewing machines had caught on. Ladies buying sewing machines from Grover and Baker in New York were shown how to use them in an elegant drawing room, richly carpeted, with rosewood chairs, tête-à-têtes, sofas, a piano, a select library, and bronze chandeliers. There was an adjoining ladies' toilet room with a looking glass, marble washstand, pins for hanging cloaks, etc. Nothing was too good for the well heeled customer.

The machines put an end to child labour in the boot and shoe trade and brought good, but inexpensive, clothing within the reach of ordinary people. But sewing machines were cheap and portable, so sweatshops could be set up almost anywhere. Meanwhile, the planters in the US South complained that no machine had yet been developed to harvest sugarcane—nobody needed one while the southern states had slaves, but that would not last.

In 1859, they hanged John Brown.

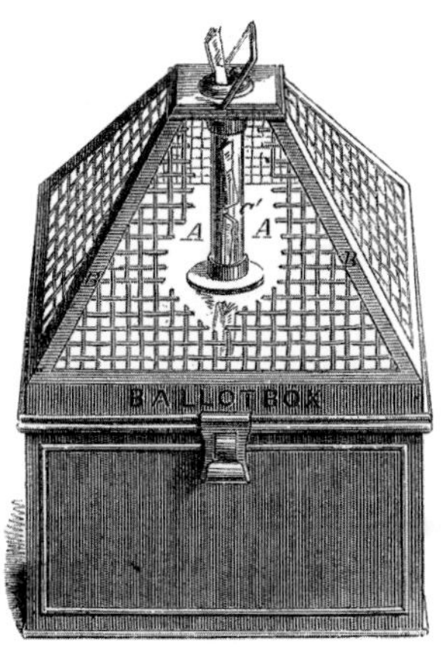


# 5: Freedom's call

*Then let us pray that come it may*
*(As come it will for a' that),*
*That Sense and Worth o'er a' the earth,*
*Shall bear the gree an a' that.*
*For a' that, an a' that,*
*It's coming yet for a' that,*
*That man to man, the world, o'er*
*Shall brithers be for a' that.*

*A Man's a Man For A' That*, Robbie Burns, 1795

Three strands of the freedom trail began in 1759. First, the defeat of the French in Canada, in 1759, meant the American colonists had no further need of a British army to protect them from French attacks, and could pursue independence. Second and third, during 1759, both William Wilberforce and Robbie Burns were born, as were Pitt the Younger, James Ruse, Mary Wollstonecraft, Georges Danton and Johann Schiller, but they don't figure here, even if each had thoughts on freedom. No coincidence is ever perfect.

It was a slow, contradictory and muddy trail. Both George Washington (married 1759) and Thomas Jefferson (a schoolboy in 1759) inherited slaves and became slave owners while championing liberty for Americans. William Wilberforce, that British enthusiast for the emancipation of slaves overseas, supported legislation at home that barred British workers from organising to obtain fair wages. Perhaps Burns was spotless, perhaps he had a skeleton of some sort in his closet as well—it matters little, because none of the other three would have worried about their attitudes at the time. By 1859, the values of decent people had changed, though the change was far from uniform, and attitudes were often apparently contradictory.

Charles Darwin detested slavery in any shape, unlike Robert FitzRoy, the captain of HMS *Beagle*, who had no objections to slavery. Yet FitzRoy aligned himself with 'Soapy Sam' Wilberforce, the Bishop of Oxford and William's son, in denouncing Darwin's explanation of evolution—for religious reasons.

Charles Kingsley, appointed as Queen Victoria's chaplain in 1859, supported Darwin, yet a few years later he used Darwin's notion of natural selection not only to justify racism in general, but in particular, to justify the controversial actions of Edward John Eyre, Governor of Jamaica, who was blamed for the brutal repression of a revolt by freed slaves on the island. In Australia, Eyre was remembered and widely respected because he had saved many lives as Protector of Aborigines at Moorundie on the Murray River in the early 1840s.

John Brown, a fervent abolitionist, preached a Christian doctrine, but was prepared to use lethal force to make his dream a reality. He raided an arsenal at Harper's Ferry, seeking arms for a slave insurrection (and to provoke a rising).

In Massachusetts, Brown was praised by many for his action, yet he was hanged for it; Victor Hugo pleaded for clemency, on the day of the hanging, saying the murder of Brown (as he called it) would be an 'uncorrectable sin' that 'would certainly shake the whole American democracy'. The *Chicago Press and Tribune* quoted Lincoln on 9 December, who addressed the South, saying:

> Old John Brown has just been executed for treason against a state. We cannot object, even though he agreed with us in thinking slavery wrong. That cannot excuse violence, bloodshed and treason. It could avail him nothing that he might think himself right. So, if constitutionally we elect a President, and therefore you undertake to destroy the Union, it will be our duty to deal with you as old John Brown has been dealt with. We shall try to do our duty. We hope and believe that in no section will a majority so act as to render such extreme measure necessary.

What this alludes to, is that the secession of the Southern states after Lincoln's election was foreseen—and not only in the North. Southern papers warned that if the candidate of the 'Black Republican party' was not overthrown in the coming presidential elections, '... it would be proof conclusive that the South could not safely remain in the present confederacy of the United States, and would have to turn to a Southern Confederacy for security and independence.' The writing was on the wall in 1859, and when Lincoln was elected the following year, a bloody war was certain.

The novel *Max Havelaar* was written in 1859. It exposed the abuses of *cultuurstelsel*, the culture or cultivation system that was close to slavery in the Dutch East Indies, modern Indonesia. It was written by a Dutch resident of Java, Eduard Douwes Dekker (under the pseudonym Multatuli), in September–October, and published in May 1860. Like Harriet Beecher Stowe's *Uncle Tom's Cabin*, serialised in 1851, it made people sit up and think. Perhaps if the equally oppressed indigo farmers of Bengal had been featured in a sympathetic novel, their lot might have been improved.

Even as help emerged for some, others suffered. In 1853, the first indentured Chinese workers reached Guyana in South America. Traders raided Easter Island in 1854, taking the population as slaves to mine Peruvian guano deposits. Later, Chinese workers were brought in, and there was a war with Spain over ownership of the guano islands.

The cocoa bean was introduced by Jose Ferreira to old sugarcane plantations on São Tome, off the coast of West Africa, in 1822. The new crop took a while to catch on. In the late 1850s, a new system began of growing cocoa beans by contracted freed slaves. It would take 50 years for the Cadbury and Quaker chocolate companies to realise they were buying the produce of those who were effectively slaves, and after this, the cruel system was abandoned. Even today, most West African cocoa comes from shady operators, using workers who are savagely exploited.

The *Clotilda*, the last known illegal slave-trading vessel to reach the US, docked in Mobile Bay, Alabama in 1860. The end was coming. The *Galveston Daily Civilian* reported on 25 January 1859 that the city fathers of Galveston had passed an anti-camel ordinance, the aftermath of an attempt to bring in slaves during 1858, with camels on board to explain away the smell of a human cargo, chained below deck under hideous conditions.

Freedom wasn't always good news. When serfdom was abolished by Alexander II in Russia in 1861, the serfs found themselves with no obligations, and with nobody owing them any obligation in return. In 1863, slavery was abolished in the Dutch colonies, but slaves were required to work another 10 years at low wages. Most decamped, leaving the plantations short of labour. Then, as now, there was a dearth of good, easy answers. As a rule, economic arguments won the day.

The first Indian coolies in St Lucia arrived under indenture on board the *Palmyra* on 6 May. In London, the British and Foreign Anti-Slavery Society protested that bringing coolies into the West Indies was only another form of slavery—even though the new Bishop of Queensland had spoken favourably of importing coolies into Australia to work the sugarcane.

After the Atlantic slave trade was largely shut down in 1807, women slaves in the US were suddenly encouraged to have and care for children, because the children of slaves became slaves themselves. Trading in slaves was profitable: two New York ladies were reported to have obtained profits in 1859 of $23,000 and $16,000 by financing illegal trading, though this may have been propaganda. In November 1858, the *Tyler Reporter* gave some sale details:

> A number of negroes, belonging to the estate of Lindsey Smith, were sold at the block in this places [sic] on Tuesday last, on a credit of twelve months, for the following high prices, viz: A girl fifteen years old, for $1,110; a woman, 22 years old, [with infant] $1,560; a girl, 9 years old, $800; a boy, 4 years old, $725.

The value of cotton produced in the US was of the order of $100 million, and this represented 75 per cent of world production at the time. Much of it was exported and, in the year ending 1 October, just over two million bales of cotton were sent to England, which relied heavily on the cotton trade.

Between 1856 and 1861, even with a depression in 1857 and a cotton famine at the end, steam power usage in English cotton mills rose from 88,000 horsepower to 282,000 horsepower. America also profited from cotton. The *New Orleans Delta* reported in early 1859 that a steamer had arrived in New Orleans from the Red River with eight widows, worth 5 million dollars. One owned 600 slaves, the others made 30,000 bales of cotton between them.

Those who could foresee the US Civil War were quick to sound warnings. There was a limited amount of land where cotton could be grown, said *Scientific American*: a belt between 30 and 35 degrees

'Human Flesh at Auction': A slave auction beneath the US flag.

on either side of the equator; in America, much of that area was unsuited to the plant, being either ocean or desert. No source for this information is given: but it was probably another shot in the war on slavery.

Britain was strongly opposed to slavery, even though Britons usually regarded indentured labour as acceptable. The US South must have feared British intervention in the looming war. This may explain why another writer pointed out that in 1850, 795 million pounds of cotton were grown in America. More than half of this went to Britain, which made an estimated profit of $187 million. The writer suggested that any temporary cessation of supply would throw hundreds of thousands into beggary—and all the landed property in the north of England would be swallowed up to maintain those who would be thrown upon the poor-rate for support.

When war did break out in the US, the cotton famine did bring poverty to northern England, in 1862–64, but despite their hardships, the textile workers were generally supporters of the Union cause, seeing the fight against slavery as a parallel to their own battles for rights and human dignity. The ideal of the Brotherhood of Man was alive and well, even if blunted.

## A SMALL EUROPEAN WAR

The Brotherhood of Man was to the fore when Marshal Haynau visited England. Were he to live now, he would be prosecuted for war crimes, for executing a number of generals after the Hungarians tried to split from the Austro-Hungarian empire in 1848. But there were other actions—particularly having women stripped and flogged (for crimes no greater than supporting the rebels)—that stuck in the craw of British workers, who dubbed him 'The Hyena'.

In 1850, the Marshal visited the Barclay and Perkins Brewery, where draymen recognised him and there are many lurid and fanciful accounts of what followed. It appears, however, that when he was identified to them the draymen used the whips they had to drive their horses, to lay into him with cries of 'Oh, this is the fellow that flogged the women, is it?' Other recorded cries were 'Down with the Austrian butcher!' and 'Give it him!' In the end, the police had to rescue Haynau and haul him away. He had more women whipped in Brescia before he died in 1858. When Garibaldi

Mildura Library

Date: 6/04/2016 3:29:10 PM

Member: Harrison, Stuart

Today's Borrowed Items:

Mr Darwin's incredible shrinking world : s
cience and technology in 1859
Due Date: 4/05/2016 23:59:00

********************************
Search, Renew, Place Holds
www.mildura.vic.gov.au/library
********************************

was in England in 1864, he visited the brewery to thank 'the men who flogged Haynau'.

In 1859, Austria's territory included a great deal of what is now Italy, and the contest over ownership led to the most bloody fighting of the decade. In the first few days of 1859, Napoleon III used warlike expressions in conversation with the Austrian ambassador, giving rise to serious apprehensions of war involving Austria and Sardinia-Piedmont. By 10 January, *The Times* Paris correspondent wrote:

> Everybody says that the war will be between Austria and Piedmont, but nobody knows who is to be the ally of Austria and who the ally of Piedmont, though everybody comprehends that Piedmont cannot carry on a war without an ally—or what will be the occasion of war, and this is precisely what one would wish to know.

In fact, with the French Emperor's comments, it was fairly clear who one of the allies would be. France was arming, and as we shall see, planning a royal alliance by marriage with Sardinia-Piedmont. By the end of 1859, the French navy would have 150 armed paddle and screw steamers of great speed, as well as sailing ships fitted with screws driven by steam. This was beside gunboats, steam transports, floating batteries and fireships.

In Sardinia, at the end of January, arms and ammunition were being stored, and assorted artillery pieces were being replaced with guns of uniform calibre. This re-arming was happening in many parts of the world, but it was more extreme in the Italian states, Austria and France. Britain was not at all sure about France, even if the two nations had been allies in the Crimean campaign. *A Napoleon's a Napoleon, for a' that,* the British rumbled to themselves, and perhaps the British had some reason to be wary.

There were those in France who were keen for a war with the perfidious English. Felice Orsini had tried to assassinate Napoleon III in early 1858. When it was revealed that Orsini had made his bombs in England, some French officers called for a retaliatory invasion of England. *The Times* was not averse to whipping up a bit of war fever, and the middle class formed 'Rifle Volunteer Corps', which attracted all sorts of people, even W.S. Gilbert, a mere clerk in 1859.

By the time he met and combined talents with Arthur Sullivan, who was studying music in Leipzig, Gilbert held the rank of captain. By the time they began writing the Savoy operas in 1875, Gilbert had probably been exposed to more than one modern major-general.

## The technology of killing

If you wanted an indication that war was changing fast, you need only look at the way patents were being granted in the 1850s. According to *Scientific American*, when '... the war with Russia broke out, the British Patent Office was inundated with belligerent projects. No less than six hundred patents have since been granted for military inventions, while the total of all that had ever been granted before was only three hundred'.

The new faster-firing breech-loading steel guns with rifled barrels loaded faster and shot more accurately over a greater range.

The English inventor and industrialist, Sir William Armstrong (he was knighted in 1859 when he took charge of the Royal Arsenal at Woolwich), had designed a breech-loading rifled gun during the Crimean War and gave the patents to the British nation. After the war, the penny-pinching government reverted to the cheaper muzzle-loaders. The Armstrong design had a screw breech, and still used shot and powder, but tests showed the Armstrong gun to be an excellent penetrator of armour and it was introduced into the British service in 1859.

Rimfire cartridges were being developed and revolvers and carbines were about to change infantry warfare. In the days of muskets, a well trained soldier could reload and fire in about 15 seconds, but an infantry charge could cover the distance between out-of-range and hand-to-hand in the time taken to fire two shots. A large and determined band could charge down a group using muskets, taking casualties on the way in, but still over-running the position. By about 1859, that principle of the infantry charge no longer held, though it took two generations for commanders to realise this.

By June, the British Army had muzzle-loading Enfield rifles, but the Royal Marines were being issued with breech-loaders, made by mass-production methods, using identical parts. The Prussian army was mostly equipped with breech-loading rifles, unlike other armies, said *Scientific American* in July. These rifles would eventually supersede all others, claimed the reporter, who thought the US might even be drawn into the war in Europe, and should start arming, just in case. It was that sort of time, that sort of world, a shrinking world where long-distance invasions were quite conceivable.

*Scientific American* noted that breech-loading guns fired three times as many shots as muzzle-loaders, that cannon now had a deadly effect at five miles, that many light field pieces were being replaced by smaller numbers firing heavier shot. Revolving carbines were being issued to light cavalry, and many-chambered pistols might be found in military holsters.

Austria and the various German armies had been profiting from the mistakes of others in armaments, but thanks to printing and faster communication, whatever one nation achieved, others soon heard about. 'Wars will be more bloody and more like murder than ever, and we hope that men may soon become convinced that it is a destructive folly, and settle their quarrels, personal and national, without recourse to slaughter and bloodshed,' warned *Scientific American*.

Even armaments manufacturing was increasingly bloody. During March, science writer Septimus Piesse visited a gunpowder works at Hounslow, researching an account that appeared in early July, in *Scientific American*. Each year, he said, the factory turned out 220 tons of gunpowder. Just 12 hours after he left the premises, the powder mills were destroyed in an explosion that killed seven people.

The armies of the 19th century were often as lethal to their troops as any enemy. When a relative of Charles Darwin, John Darwin Wedgwood, gent. (short for 'gentleman'), entered the 61st Foot as an ensign in early June, he purchased his commission, a practice that only stopped in 1871. An officer's life was not always easy, as a number of melodramatic novels suggest, but this December advertisement in *The Times* shows how real the problems were:

> A young officer wants a loan of £300 to save him from ruin. He cannot offer any security, except insuring his life for the amount, and an honourable determination to repay it by instalments of £50 per annum. He makes this appeal in hopes that it may catch the eye of some noble-minded lady or gentleman who may have more money than they want. Address, M.N., post-office, Carlow, Ireland.

Candidates for army commissions needed no aptitude. Each had only to offer proof of age, a certificate from a minister of a church that he had been properly instructed in the principles of religion, and a certificate of a master or tutor that he had been of sound moral character for two years. And then take examinations. Candidates needed a total of 1800 marks, to be obtained in various examinations in classics (3600), mathematics (3600), English (1200), foreign languages including French and 'Hindostani' (1200), natural sciences, mineralogy and geology (1200), experimental sciences (chemistry, heat, electricity including magnetism, 1200) and drawing (600).

Mathematics and English were obligatory. Marks gained in the optional subjects would not count unless at least one-sixth of the marks were gained. Mathematics included fractions, proportion, square roots and simple interest, simple algebraic equations and the first three books of Euclid. The pass mark was 400/1200, with at least 200 marks to be obtained in arithmetic. To enter the staff college, officers had to pass compulsory examinations in mathematics, fortification, military topography, military law, military art and history and French. Voluntary subjects were

German, Hindustani, Chymistry (a not unusual spelling in 1859) and Geology.

With such an officer selection process and ever more deadly weapons becoming available, it is little wonder that troops were so often slaughtered with no gains being achieved. When *Scientific American* reported the end of the war between France and Austria, the writer commented that Solferino was won 'by the bayonet', as was Inkerman in the Crimea. The war had ended in '... so many more useful lives swept away, or, as the tyrants think, so much more of the rabble killed'. Clearly, the writer concluded, the modern army needed artillery able to carry ball long distances with perfect aim, rifles—and bayonets.

Warfare may have seemed well run to Victor Emmanuel II of Piedmont-Sardinia, Franz Josef of Austria and Napoleon III of France, who looked on from a distance. But those who saw from close quarters may have seen differently. One such man, Henri Dunant, a Swiss citizen, witnessed what happened when these three royal gentlemen with no particular military training, prowess or track record, took the field to play a game of chess with live pieces and live munitions.

That sort of game is damaging to life and limb. Like others, Dunant saw the wounded who had later been slaughtered where they lay, and he heard the surviving wounded and dying crying out for water. Horrified, Dunant decided to do something, to push for change.

One good thing did emerge from Solferino. It took five years, but in 1864, what we today call the International Red Cross, was formed. It also led to the first of the Geneva Conventions.

Assorted schemes needed to be in place to start a war and, later in the chapter, we will examine the key step in one of these manoeuvres. Napoleon III and Cavour, Piedmont-Sardinia's Prime Minister, agreed in 1858 that the time was right for a Northern Italian kingdom to be established.

Napoleon undertook that if Austria attacked Piedmont-Sardinia, France would join the war on Piedmont-Sardinia's side. Cavour immediately set to work to provoke Austria into starting a war. The Piedmont-Sardinia army was mobilised, Austria mobilised in response, and then served an ultimatum on Piedmont-Sardinia to demobilise, and in April 1859, crossed the Ticino River. This made Austria the nominal aggressor.

The French promptly joined in, winning two costly victories at Magenta and Solferino. Losing his nerve or his appetite for battle, Napoleon III then withdrew from the war with the Truce of Villafranca. The treaty that followed left Venezia in Austrian hands, but gave Lombardy to Piedmont.

Other Italian territories were returned to their former rulers who had been ousted by the Italian people, but by March 1860, plebiscites had transferred Tuscany, Modena, Parma, Bologna and Romagna to Piedmont-Sardinia. Napoleon III graciously accepted these variations, in return for France getting Savoy and Nice.

The Russians conquered the North Caucasus in 1859, opening the way for them to exile a number of groups, including the Chechens, to Anatolia in 1877–78. Spain was making war in Morocco, the French were fighting in Algeria and attacking Vietnam, while French and English forces took up a joint offensive against Chinese wickedness. Chinese defenders had fired on Royal

The Battle of Solferino, 24 June 1859.

Navy ships and two French gunboats which were trying to force a timber boom across the Peiho River. Said the *Moniteur* in Paris: 'The Government of the Emperor and that of Great Britain are about to take measures together to inflict chastisement and obtain every satisfaction which so flagrant an act of treachery requires'.

The Chinese had agreed, under duress, to open the Yangtze River and extra ports to the British, but when, in 1859, the Chinese 'violated' the provisions of the treaty of Tientsin (Tianjin), fighting began again. British forces entered Peking (Beijing) in October. The commodity exchange in Singapore listed opium at $385 per chest: in 1859, it was a legally traded substance, not only because it was a part of Perry Davis' Pain Killer, but because Britain was draining its gold reserves to pay for tea and needed a trade item to sell to China to balance the tea imports. Britain had decided that like it or not, China would buy opium sourced from British India. It was a solution with all the fairness of a modern Free Trade Agreement.

## NATIONAL PRIDE

Between 1857 and 1861, Mexico was wracked by a bloody civil war between conservatives and liberals, the War of Reform. Early in 1859, a joint intervention in Mexico by British, French and American forces seemed likely, but cooperation was never certain. In the end, the US stood back when Britain, France and Spain landed troops in Mexico in 1861, resulting in the imposition of Maximilian of Austria as emperor of Mexico.

Just as enemies could become allies overnight, even the best of friends could quickly fall out. Later in the year, British and American troops almost started a war between their nations in a strange event known as the 'San Juan Island Pig War'. At stake was the possession of one island near Vancouver, about 55 square miles (140 square kilometres) of land. The sole casualty was a British pig, killed by Americans on 15 June when it attacked an American garden.

Anything could lead to war, and many people seemed to want it. In November, authorities in the Canary Islands were in a lather over a book written by the wife of the British consul, a work which reflected badly on the residents. *The Times* said her comments were no harsher than those of the local people, but the book was banned, and consul Murray's dismissal and removal were demanded. Spanish honour required no less.

Beyond, and sometimes counter to, nationalism, people around the world were beginning to claim their right to political, religious, economic and cultural independence. The ideas had probably always been there, but now rail and steamship transport could move organisers (or agitators, depending on your point of view) from place to place. There were also mails to pass ideas on, and more printing presses. In January, Wallachia and Moldova, two Ottoman principalities, were unified as Romania. First *The Times* reported that 'Prince Michael Stourdza, who is enormously rich, is likely to be elected Hospodar of Moldavia. Prince Stirbey, who has the best chance in Wallachia, is still at Crajova, in Little Wallachia'.

The people decided otherwise. Even before the story appeared, Moldavians elected Colonel Alexandru Ioan Cuza as their leader on 5 January. On 24 January, Wallachia elected him as well. The Ottomans agreed to this on the condition that Cuza visited Istanbul and rendered homage to the Sultan. Greece had also been hived off from Turkish control and the Magyars, among others, hoped to be freed from Austrian domination, especially after Austria let go of Lombardy and, to other eyes, showed signs of weakness. In Ireland, the Irish Republican Brotherhood (IRB) was formed by James Stephens and John O'Mahoney around 1858–59. Known also as the Fenians, the IRB was the forerunner of the Irish Republican Army. At this time, Paul Cullen recruited Major Myles O'Reilly to form an Irish Brigade to defend the Papal States against Garibaldi. As a reward, Cullen was made a Cardinal in 1866 by Pius IX.

Some freedom movements took longer to gain a foothold. The people of India, like the mass of the British populace, remained almost powerless, but around the world, the conditions of ordinary people were slowly improving.

## WINNING THE VOTE

In 1859, just one in 30 of the English population had the vote in Britain. In the established colonies of Australia, however, in the rich and gold-soaked land, all adult male British citizens had the vote, even then. Giving the vote to all white males was a step in the right direction; giving the vote for women would take more time; giving the vote to Aborigines would take longer still.

The 1832 *Reform Act* had increased the British electorate from about 500,000 to 813,000. At the start of 1867, there were one million voters. After the reforms of that year this doubled to two million, and the *Reform Act* of 1884 raised it to about six million. Changes in 1918 gave the vote to men over 21 and women over 30, with women over 21 getting the vote in 1928. Today, the free world takes it for granted that we should have electorates of equal size, universal suffrage, secret ballot, pay for representatives and the right of all voters to stand for election. These were all Chartist policies, requested in the 19th century. It is only their unworkable demand for annual elections that has never been implemented.

During 1859, the Chartists faded, winning a mere handful of votes in the 1859 election in Britain and they were officially wound up in 1860. Their democratic ideals, however, were taken up by the new Liberal Party. Chartism had far more effect in Australia, where the gold diggers' insurrection at the Eureka Stockade in 1854 had seen an overtly Chartist set of demands presented to the government. In the absence of an established power bloc, the government was forced to cave in.

In Britain, there was a fear of giving the vote to 'the mob'. Prince Albert, usually a man of sense, had called the beating of Marshal Haynau 'a slight foretaste of what an unregulated mass of illiterate people is capable' and he was not alone. During the year, John Stuart Mill published *On Liberty*, and wrote '... no lover of improvement can desire that the *predominant* power should be turned over to persons in the mental and moral conditions of the English working class'. The Melbourne correspondent of *The Times* wrote:

> It would be impossible at this time to anticipate what is likely to be the effect of what is called manhood suffrage on the general elections, which are expected to take place in August, for, although this radical change was introduced by Mr. Haines in November, 1857, it has not hitherto had an opportunity of displaying its operation.

In Britain, where there was no secret ballot, the general election of 1859 led to hearings about bribery of voters, from which nine members were found guilty of misconduct and unseated. Members had moved into the new Parliament House in 1857 (the old one burned down in 1834), and in the last days of 1858, the great bell of Westminster, known as Big Ben, was swung in the clock tower of the Palace of Westminster and tested just before Christmas, ready to be activated in the New Year. Another stone had been laid in the wall of British imperial grandeur, and the lower classes knew little of the corruption beneath the glitter.

## A QUESTION OF CLASS

In London, May, June and July were once the months when Parliament met, and this determined 'the season' which ended on 12 August, the first day of grouse shooting. The well off, even those not involved in politics, came to London in the season for races at Ascot, opera, balls, parties, viewings of the Royal Academy and other social events.

Every member of society had obligations, but the obligations of some were less onerous than the demands society made of others. In an age before labour-saving devices, the rich had a duty to hire labour. Mrs Beeton began publishing the articles that would form her *Book of Household Management* in 1859, and in these, she even listed how many servants a household should have, based on income. Everybody in society had their obligations: army officers, having purchased their commissions, were obliged to lead, and sailors and soldiers were obliged to submit to flogging.

Still, the writing was on the wall for the lash as a naval punishment after an incident at Plymouth in July, on board HMS *Caesar* (then in dock) when a sailor was flogged in front of a large number of artisans. The outraged watchers protested loudly and argued with some of the officers. Flogging was not abolished in the British army and navy until 1881, but it effectively ended in 1859, largely thanks to one crusading doctor.

When Mary Ann Evans published *Adam Bede* in 1859, she wrote as George Eliot, and was unprepared for the attention it would bring her. Well known and admired in her own circle of intellectuals, she now found herself publicly identified with George Eliot, the clever 'male' novelist. By the time *Middlemarch* came out in 1871–72, she was well respected, and admired, in both names, for her social conscience.

In chapter 16 of *Middlemarch*, her fictional characters debate Wakley's view that coroners need medical training, so as not to be bamboozled or misled by inadequate medical men. Unlike the characters in the book, Thomas Wakley was a real person. As a young doctor he cared about the reform of the medical profession, and political reformer William Cobbett suggested that he establish a medical journal, which he did in 1832, calling it *The Lancet*.

In 1835, Wakley was elected to Parliament, and his maiden speech attacked the conviction of the Tolpuddle Martyrs, a group of unionists transported to New South Wales in 1834 on trumped-up charges for daring to organise to defend themselves. He was an all-round decent human being of liberal outlook who opposed slavery, the Corn Laws, the 1834 Poor Law and the *Newspaper Stamp Act*. Wakley deserves most of the credit for the creation of the Royal College of Surgeons in 1843 and also the General Council of Medical Education and Registration in 1858, but his public fame rests mainly on his work as a coroner. He not only held the beliefs mentioned in *Middlemarch*, he put them into action.

No informed observer of coronial inquests today can hear a coroner's blunt demolition of an evasive witness without recalling what happened when Wakley confronted a workhouse master. The man complained that an exhumed pauper's body, while it had undoubtedly been scalded to death, had not been properly identified as Thomas Austin, the subject of the inquest. Said the worthy coroner: 'If this is not the body of the man who was killed in your vat, pray, Sir, how many paupers have you boiled?'

Before he could make his name as a coroner Wakley needed to be elected to the post. He narrowly lost his first attempt in East Middlesex in 1830, but won in West Middlesex in 1839. From time to time, Wakley reported details of notable coronial hearings in *The Lancet*, and that brings us to his inquest into the death of Fred White, a young soldier of the Queen's Own Hussars who died in 1846. The true cause of death, a flogging of 150 lashes, was covered up, but with a jury's support, Wakley ordered the body exhumed so the original post mortem could be assessed. The evidence of military cruelty was there for all to see, and an end to flogging in the army came a step closer. The practice did, finally, cease after a man called Davies was flogged almost to death at Woolwich in September 1859; Wakley's pursuit of the White case had laid the foundations.

Wakley and Dickens met in 1841, and Dickens once served on a jury under him. The two undoubtedly influenced each other, but Wakley's essential humanity is seen best in an instance where he may have cut the odd coronial corner. Thomas Glover was a Civil Surgeon at Scutari during the Crimean War, and either during that time or on his return, became addicted to chloroform and opium, both then readily available to doctors. In April, 1859, he died as a result of an excessive dose of chloroform, and his colleague Thomas Wakley, acting as coroner, brought in a verdict of accidental death. It did no harm. But there was harm enough around.

## EXPLOITING THE WORKERS

Even that sainted champion of the working class, Karl Marx, was not above seducing his wife's maid, who bore him an illegitimate son. Small wonder, then, that so many educated and rational people managed to avoid seeing the ugly horror that was around them. It was the same with war: until the telegraph and the first war correspondents, people could ignore what happened on the battlefield and let poets emphasise the benefits. The Light Brigade rode to their deaths but Tennyson made them the 'noble six hundred', and all was well.

Like bayoneted soldiers dying of thirst on the battlefield at Solferino, ordinary workers knew all too well what it was like to have your life ebb away, but they could also see their children dying of disease or starvation. In their desperate plight, they could not see a need for any national aspirations, but they nurtured hopes of their children's survival, and they knew well what was needed. Mr Darwin saw the natural struggle to survive as a tool of evolution, but struggle was a fact of life for the poor.

Mrs Beeton said that with an income of £1000, a family should have five servants. The lowest-paid would get perhaps £10 or £12 a year, a place to sleep and cast-off clothing, while a housekeeper might get £25 plus tea and sugar, and a butler in a large household could expect as much as £50 a year plus keep. From other sources, a minister of religion might expect £150, an Australian postmaster received £200 and an important official, like Richard O'Connor, clerk of the New South Wales Legislative Assembly, had £800 a year in 1856.

Depending on supply and demand, the average labouring adult received around thirty shillings a week for a six-day week, some £75 a year. In San Francisco, skilled masons earned $10 (£2) a day, paid daily in gold. Those were long days, generally ten hours or more, and hours attracted more attention than the pay.

On 1 January, an advertisement in *The Times* offered a prize of five pounds for the best essay of less than 32 pages, arguing for the shortening of the hours of labour to nine hours '... as at present agitated by the Building Trade'. In Scotland, the miners of West Fife supported a petition to Parliament calling for a restriction of the working hours in pits to eight or nine hours a day. They also sought a shortening of the hours of labour for those under the age of 14 years. The petition asked that mine owners should be obliged to provide school instruction for all young persons under that age employed in the pits; and that the owners should be obliged to adopt the safety cage and other scientific appliances 'to prevent the sacrifice of human life'.

The building trade took the lead on hours. By August, the Nine Hours Movement had 20,000 skilled artisans in the Confederation of Building Trades, and began admitting labourers. If the artisans' hours were reduced, those of labourers would be as well, so they were welcomed as brothers and equals.

The employers had financial reserves and felt they could hold out during a long break with no income, so a vicious attack began when a firm called Trollope and Sons dismissed a mason for presenting a proposal for a nine-hour day. The workers struck, demanding both a nine-hour day and the man's reinstatement. Instead, the other major builders closed their projects in sympathy, throwing the rest of the building workers out of work. There would, the owners declared, be no re-opening until Trollopes' workers went back, and they further demanded that workers sign a pledge 'not to belong to any society which interferes with individual trade arrangements'.

This pledge was rejected as an 'obnoxious document', and other unions promised financial support. According to *Scientific American* in early September, '[p]ublic opinion in London and the whole kingdom appears to be on the side of the operatives, and it is believed that a compromise will soon be effected between them and their employers ...' As you might expect, Karl Marx was in fine form in *Das Volk*:

> The obstinacy of the masters, who lay claim to an authority over their 'hands' similar to that of the American plantation owner over his slaves, has evoked the disapproval even of a section of the bourgeois reporters ... Generous money contributions are streaming into the 'society' from all parts of the country, but so far the unemployed workers have refused to draw on these. Honour to the brave! Would the bourgeoisie be capable of such sacrifice for the sake of its class interests?

In February 1860, workers withdrew their nine-hour demand, and the 'document' was quietly dropped. The owners had maintained the status quo, but the workers had survived, and withstood an attempt by the employers to force them to agree not to join a union.

That said, the unions and their supporters were not entirely flawless. A mason was imprisoned for two months after he intimidated workmen during the builders' strike, and an attempt was made to blow up a house in Sheffield in which a Mr Linley lived, apparently because he refused to join the saw grinders' union.

Britain's *Molestation of Workmen Act* of 1859 allowed that any person, employed or not, could enter into an agreement with others to fix a wage rate and hours. Moreover, it also specified that they were not liable for prosecution just for trying peaceably to persuade others to strike, since the act defined this as neither molestation nor obstruction. There would be no more like the Tolpuddle Martyrs. Slowly, working people began to win some freedom.

## THE OTHER HALF

Nations liked having royalty to rule them, and would desperately seek out spare royalty to meet local shortages. Sweden even took Bernadotte, a former French revolutionary and marshal of France (but a commoner), as their crown prince in 1810, Greece took Otto of Bavaria as king, until 1862, when he was replaced by a Danish princeling, and Mexico had Maximilian of Austria foisted on them as their emperor—until they shot him.

The first Maori king was appointed in New Zealand, around June 1858 (or 1859 according to some, but certainly no later). It was a symbolic role that continues to this day. Meanwhile, Victoria's daughters were marrying into the royal houses of Europe, spreading the genes of haemophilia far and wide.

Thanks in no small part to the efforts of Prince Albert, her hard-working and intelligent husband, named 'prince consort' in 1857, the royal family was not the mess it had been when Victoria was born. Albert bought Balmoral in 1852, having rented it since 1848; this pleased the Scots. But he also contributed by cultivating the scientists of the day and encouraging technological progress. The Queen ruled, but Albert looked after the hearts and minds.

The Queen was also forward looking, though Dr Wakley was displeased in 1853 when she allowed Dr John Snow to administer chloroform to her during the birth of Prince Leopold. Snow was not singled out by name, and Wakley's *Lancet*'s comments were careful to suggest only that the Queen had received bad counsel. Nonetheless, the sovereign accepted chloroform again at the birth of her last child, Princess Beatrice, in 1857.

In April 1859, a Juvenile Fancy Dress Ball was held at Buckingham Palace to celebrate Prince Leopold's sixth birthday. Leopold, the second-last of Victoria and Albert's children, was a haemophiliac and carefully protected, though he seems to have had a happy enough boyhood. In later years, he struggled to escape his mother's concerned care, studying at Oxford and enjoying the company of art critic John Ruskin and Professor Max Muller, one of the leading linguists of the era. He also found pleasure in the company of academics' families, including that of

Dean Henry George Liddell, better recalled today as the father of Alice Liddell, who inspired Lewis Carroll to write the two 'Alice' books.

Besides the genuine royalty, there were the ersatz royals, eccentrics like Californian Joshua A. Norton who proclaimed himself 'Emperor of These United States' in 1859. Norton had lost his money when an ambitious scheme to corner the San Francisco rice market failed. He lost his reason as well, but he was accepted by the folk of San Francisco—who humoured and even encouraged him—with one tale reported of him placing himself between white race rioters and their Chinese targets, and persuading the rioters to desist. It made him one of the more useful specimens of royalty.

The impressively named Marshal-General George Henry de Strabolgie Neville Plantagenet-Harrison was born plain George Harrison in 1817. He had served with distinction as a soldier in several South American armies, in Denmark, and in Germany, though his rank of marshal-general came from his time with the army of Corrientes province in the Argentine Republic. He claimed to be a direct descendant of John of Gaunt, and so the rightful ruler of the British realm.

Harrison later returned to England, but from 1850 was refused access to the British Museum. He said this was because he claimed to be Duke of Lancaster, though museum authorities may have had a simpler reason, since he appears to have been more than a little eccentric. In 1858, he tried, unsuccessfully, to get himself summoned to the House of Lords as Duke of Lancaster; he was in the Insolvent Debtors' Court in March of 1859; and, in 1861, he was declared bankrupt and confined in the Queen's Bench prison.

## The art of starting a war

The official royals could still influence history. At the start of 1859, the capitals of Europe were reminded that Princess Clotilde, the eldest daughter of King Victor Emmanuel of Sardinia, had been born 2 March 1843, and would soon be 16 years of age, and marriageable. It was expected she would be betrothed to Prince Napoleon, a nephew of the French Emperor.

A report in *The Times* on 27 January recorded that General Niel, ADC to the French Emperor, had made a formal demand in marriage of Princess Clotilde, which was graciously acceded to. No time was wasted and, on 4 February, *The Times* reported 'Prince Napoleon and the Princess Clotilde arrived at the Tuileries at half-past 3 o'clock this afternoon'.

After that, the mystery of who would side with whom in the coming war between Austria and Sardinia was a mystery no more. One of the shocks of 1914 was that Britain should be at war with Germany when it was ruled by Queen Victoria's grandson. By 1914, the rules had changed, but in 1859, policy was still arranged by royal alliances as much as by economic or military needs, though at times, young royals were required to enter alliances that reflected those needs.

Still, if occasional harsh requirements faced young blue bloods, it remained an era of smug superiority: aristocrats over commoners, rich over poor and white over other races, with many finer gradations to be found in each of those scales. It was a superiority which conferred the right to repress, to exploit and to colonise—and, sometimes, all three at once.

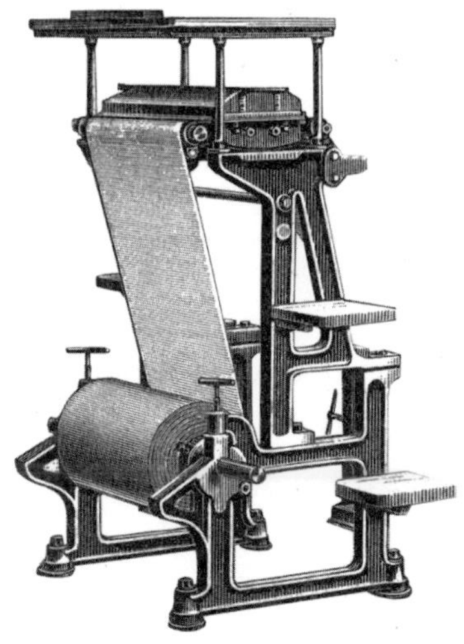

# 6:

# To rise in the world

*Richard Arkwright . . . was never at school: the only education he received he gave to himself; and to the last he was only able to write with difficulty.*

Samuel Smiles, *Self-Help*, 1859.

The well-off knew the way out of poverty: 'improve yourself', they told the poor. Samuel Smiles edited the *Leeds Times* from 1838 to 1842 and he also lectured on self education at working class institutions in Leeds. His talks became the basis of his 1859 book *Self-Help* which sold a healthy 270,000 copies all over the English-speaking world before he died in 1904. A quick worker, Smiles also published a life of George Stephenson in 1859, taking advantage of Stephenson's death to celebrate the engineer's achievements.

The Ragged Schools Union in Britain was formed in 1844 while the Sunday School Society was founded rather earlier in 1785. By 1859, the hunger for learning was far greater than the two organisations could meet. In 1851, the census showed 1545 evening schools in England and Wales, all open to men who were over 16 and competent in reading, writing and basic arithmetic. Later, elementary classes were added and women were admitted. There were also about 23,000 Sunday schools in England with almost 2.1 million children enrolled. Except in London which had more patchy coverage, the Sunday schools probably reached most of the children of the working class.

Most Western countries aimed to provide education, or at least technical training, to those who wanted it. The names varied (in Australia, the venues were generally called Mechanics' Institutes) but they all had similar aims of improving the poor. Factories in England and Scotland often provided schools at the manufacturers' expense to educate the employees. American cities offered night schools, but these were only supported in winter and so were less effective.

Mindless education-by-lowest-common-denominator in the schools was pilloried by Charles Dickens when he depicted Mr Gradgrind and Mr Squeers. But such characters were all too real, as two advertisements in *The Times* of 27 January demonstrate. Kingsdon House near Harlow offered:

> EDUCATION Backward pupils rapidly brought forward in all the requisites of a sound commercial education. The locality is unsurpassed. Terms reasonable.

Another advertiser sounds even more terrifying:

> UNRULY CHILDREN.—Ladies and gentlemen who find their sons and daughters getting beyond their own control are recommended to apply for advice and assistance to a married clergyman, of much experience in gaining the esteem and confidence of young persons of all ages from 5 to 20. Address A.M. at H. and C.'s, general advertisers, 13 Lombard-street, E.C.

Even soldiers were to be educated. A number of lectures were planned, early in 1859, for the soldiers in winter camp at Shorncliffe. Some would be delivered by army schoolmasters, though Mr Mackinson of Hythe was to lecture on 'The Geology of our Coast'.

In London, the Royal Institution offered evening lectures on a Friday. In the afternoons, Professor Owen spoke about mammals (on Tuesdays), and Professor Tyndall expounded on the force of gravity (on Thursdays).

It had been normal for the colonies to send their young men to Britain to be trained at university, but by 1850, Canada had universities in Québec, Ottawa and Toronto. By 1859, Australia had universities in Sydney and Melbourne, with the Gothic Revival Great Hall of the University of Sydney opening on 18 July. There may have been empurpled bush to the north and west of Sydney University where it lay on the outskirts of a small town in 1859 (it is now entirely hemmed-in by urban development), but the colonies were improving themselves and would now deliver higher education to more of their young men (and in due course, to selected young women).

Even in Australian country towns, education moved ahead. In 1859, the local member of parliament had land set aside at 93 Brisbane Street, Tamworth, for the building of a Mechanics' Institute, and the colonial government provided a grant of £150 towards the cost. The building opened in 1866 and, even if it is now the Australian Country Music Hall of Fame, over the years, a number of rural minds were improved there.

Oxford was also in ferment in 1859. New Chairs in Logic and Moral Philosophy were created, the start of a move away from each college tutoring its own undergraduates. A law school had been established there in 1850 (and at Cambridge in 1858), and medical education was turned on its head and beaten into shape in the late 1850s. School education in Britain and the colonies was slower to improve, but the basic intentions were in place in 1859. People might not be doing much, but they were talking about change.

## LITERATURE AND LITERACY

Better education meant more people eager for something to read, while better lighting and shorter working hours gave them more opportunities to read. The late 1850s and the early 1860s saw many new journals while *Scientific American* rejigged its format and started a new numbering for volumes.

Other titles changed hands: the bidding for the copyright of *Household Words* began at £500, but by £1100, only Charles Dickens and Messrs Bradbury and Evans were in the race, reported *The Times*. Rights to the title were eventually won by Dickens, for £3650, but Bradbury and Evans immediately started *Once a Week*. *Macmillan's Magazine* began in 1859, and 1860 saw *Cornhill Magazine* and *Temple Bar* offered for sale to the reading public. Each offered steady cash incomes to writers, and an exposure to new readers.

In a purely symbolic sense, 1859 was a year of literary change. Sholom Aleichem, Arthur Conan Doyle, Kenneth Grahame, A.E. Housman, Fergus Hume and Jerome K. Jerome, all literary innovators in later years, were born in 1859, as was Havelock Ellis. More to the point, in the September issue of the *Gentleman's Magazine*, two reviews appear, one after the other: Washington Irving's *Life of George Washington* and *Selections Grave and Gay* by Thomas de Quincey.

By the end of the year, both were dead. So were Leigh Hunt and Thomas Macaulay, but most years would probably offer a similar harvest of arrivals and departures.

If the rise of journals helps define the period, a list of the publications of 1859 tells us more. The year saw Walt Whitman's *Out of the Cradle, Endlessly Rocking*, Dickens' *A Tale of Two Cities*, published in parts from April to November to promote sales of *All the Year Round*, Darwin's *On the Origin of Species*, Dinah Craik's *A Life For A Life*, about the reform of a man who has committed murder and a woman who has had an illegitimate child, the first edition of FitzGerald's *The Rubaiyat of Omar Khayyam* (severely revised later), Thackeray's *The Virginians*, and the start of the twelve-cycle poem, *The Idylls of the King*, by Tennyson.

Detective fiction probably has its roots in 1841, when Poe's *The Murders in the Rue Morgue* came out, though the Newgate Calendar had long offered 'true crime' with fictional embellishments. Dickens had toured the London slums in 1842 with a London policeman, Inspector Charles Field, and in 1853 he based Inspector Bucket of *Bleak House* on Field. By 1859, Field had retired but was undertaking private enquiries, offering a factual basis for detective fiction.

At the end of 1859, Wilkie Collins' *The Woman in White* began to appear in *All the Year Round* in England and *Harper's* in the US and as a book in 1860. The work set a standard for later mystery novels and had its origins in a real incident, but Collins also took details from a French crime case to flesh it out. Two of 1859's new babies, Arthur Conan Doyle and Fergus Hume, would extend the genre.

John Lang was an Australian who was either remarkably prolific or who had arrived in London with a number of manuscripts. In 1859, he published *My Friend's Wife, The Secret Police, Botany Bay or True Stories of the Early Days of Australia* and *Wanderings in India and other Sketches* (first published in *Household Words*), also in 1859.

Thomas Hughes published *Tom Brown's Schooldays* in 1857 and *Tom Brown at Oxford* in 1861, bracketing 1859. In 1865, Hughes was the first candidate elected to the British Parliament with trade union support. He ran Maurice's Working Men's College from 1854, a neat example of the way literacy, education and self improvement were intertwined. Hughes was writing a new style of children's story, at a time when most reading material for children was ghastly. Frederick W. Farrar's *Eric, or Little by Little* was a last shudder of the old earnest improving style when it came out in 1858, but even a work like *Alice's Adventures in Wonderland* (1865) had a hidden agenda of teaching logical thinking through tightly constructed nonsense.

Reading for adults was also often intended to improve the mind by heaping high the spoons of sickly saccharine. In late July, *The Times* featured an advertisement for '*Helen Lindsay; or the Trial of Faith,* by a clergyman's daughter', which quoted several reviews. *The Athenaeum* said: '*Helen Lindsay* will be read approvingly in family circles where works of fiction do not generally meet with a cordial reception'. *The Observer* said: 'It deserves to be highly appreciated, so conscientious and so pure is its aim and object'.

Luckily, George Eliot was on hand to deliver a restorative dose of *Adam Bede* to an ailing literature. *Adam Bede* and *The Mill on the Floss* after it, gave prestige to the novel that featured dialect, while the Burns centenary in 1859 reminded people of another sort of dialect.

William Barnes had published his *Hwomely Rhymes* in 1858, offering verse in Dorset dialect; Charles Kingsley used dialect in *Westward Ho!* in 1855, and Charles Dickens had presented Sam

A Parisian primary school room, circa 1859.

Weller in dialect for comic effect even earlier. There was something in the air.

It seems appropriate that it was January 1859 when the Philological Society issued their 'Proposal for the publication of a New English Dictionary'. In it, they called upon the English and American public to help collect the raw materials in the form of quotations from writers of all ages. The result would eventually be the *Oxford English Dictionary*.

In the US, Messrs G and C Merriam were about to publish a dictionary with 1500 illustrations. The pictures were described as 'well drawn, well engraved and very truthful, and which will impress their definition upon their [young folks] minds as the light daguerreotypes the image upon the metal plate'. The second edition of John Bartlett's *Dictionary of Americanisms* was also published in 1859.

Thomas Wade published his *Peking Syllabary,* a way of consistently rendering Chinese, where the syllable words have tonal values that alter their meaning, into a Roman alphabet. This allowed those who did not know Chinese characters to recognise, at the very least, personal names and place names. While this may not have helped shrink the world, it certainly brought East and West closer together. When he retired in 1883, Wade donated his Sinology collection to Cambridge University, and became the university's first Professor of Chinese, retaining access to his own library.

## The role of the library

Libraries were important then, just as they are now. An 1859 census revealed 50,000 libraries in the US holding 12 million volumes. This makes an average stock just 240 volumes, but the list included 18,000 libraries of common schools and 30,000 libraries of Sunday schools, all of them on the small side.

From the records of the Athenaeum Library, the most popular author in the US was Sir Walter Scott, followed by both familiar and unfamiliar names: Simms, Cooper and Dickens, almost equal, then Washington Irving, Mrs Stowe, 'Prescott the historian' (probably William Hickling Prescott, who died in 1859), Charlotte Bronte and a Mrs Henzt, then Bulwer, Longfellow, Willis, Kingsley, Thackeray, Abbott, Macaulay, James, Bayard, Taylor, Curtis, Hawthorne and Bancroft (the same Bancroft who travelled to San Francisco in 1852, who founded that city's Bancroft Library in 1859).

Shakespeare was low on the list but observers said this was probably because most people owned a copy. In Britain, a subscription to Mudie's library was one guinea, with country subscriptions two guineas and upwards, depending on the number of volumes required. Mudie's circulating library added 200,000 volumes in 1858–59, and was increasing the collection at the rate of 120,000 volumes per year.

On the other hand, the Bannatyne Club, instituted in 1823 by Sir Walter Scott to publish works of Scottish poetry, literature and history, was dissolved in Burns' centenary year. Perhaps they thought it had done all that was required of it, or perhaps people were pursuing new and more modern interests, like photography.

## PHOTOGRAPHY AND ART

A good pencil artist could, and still can, capture an excellent likeness in twenty minutes, but photography could do the same thing, and much faster. It needed training in manipulation, but the effort required was far less than that needed to become a good artist, and talent and aptitude ceased to be a factor, if one sought only a factual record.

William Makepeace Thackeray trained as an art student from 1833 to 1837 and when he realised his technique was poor, he turned to journalism, and later became an art critic. He was appointed editor of *Cornhill* magazine in 1859, but it was George Sala, in *Household Words*, who drew readers' attention to the rise of photography and the passing of the painter. Sala advised that the police were using photos of criminals, and that some had even suggested attaching photographs to the passports of foreigners. Along with railways and gutta percha, photography was one of the 'commonplace marvels' of his time, he said.

In September, *Scientific American* featured an engraving of Charles Goodyear labelled '[From a Photograph.]', but usually, this sort of help was not acknowledged. The most common portrait of Charles Dickens is the one which appeared in the *Illustrated London News*, 18 June, 1870. It shows a pose which tells us the engraving was based on a photograph, given the way Dickens' head is propped by his right arm. His left leg is crossed over the right, and his left arm rests on the arm of a chair while his left hand is on his left knee, a perfect pose to immobilise the subject during a slow exposure.

*Scientific American* declared at the start of the year that photographic engraving was advancing and would soon advance more. During 1859, Firmin Gillot developed a new method for etching metal plates, but the progress continued. In 1872, his son invented zincography, combining photography with etching so the resulting picture could be sized up and down, and by 1880 a method of producing intermediate tones was devised by using a system of dots of different sizes. From a simple start in 1859, any publication could then offer illustrations, with no need for expensive artists.

William Frith's *Derby Day* was painted in 1858 and caused a near riot when it went on display, such was the public's enthusiasm for it. Crush barriers, and a policeman on guard, were needed to manage the enthusiastic crowds. Nobody mentioned that the amazing images had been captured in part from a series of photographs, taken on Derby day for the artist's use. Most artists continued to paint or draw from life—or death. Dickens described an interruption which arose while he was dining with Sir Edwin Landseer, who sculpted the four lions at the base of Nelson's column. During the meal, a servant entered to ask 'Did you order a lion, sir?'

The records show Landseer ordered at least two lions at various times. He bought a newly dead lion from a menagerie and dissected it until the neighbours complained of the smell, and once when the Duchess of Abercorn called upon Landseer, she found him shaping a huge mass of clay while an elderly and docile lion wandered around the garden.

The way Dickens was required to pose for his photograph helps to explain why painters of animals were able to flourish longer than landscape artists and portrait painters. Animals simply would not hold still long enough to be captured on film, while artists knew how to work around any restiveness.

If a street scene had to be photographed, the trick was to get up early on a summer Sunday morning, and then persuade the few folk in view to hold still while the shot was taken. In time, asking for stillness ceased to be an issue as film became more sensitive and, therefore, quicker to use, but for a considerable time, only painters could deliver images with colour.

James Ward RA died at the age of 90 in 1859 after a lifetime as an animal specialist. As a young man, he painted a cow, and went on to paint more than 200 further cattle portraits in a career which introduced him to all the leading figures in the 'cow appreciation world'. He also painted horses, but mainly did cows in a life that brought him much satisfaction.

Ward was born in 1769 near Dowgate Hill, by then the home, for almost a century, of the dung hill where horse manure was taken after it was collected from London's streets, along with cow manure from the city's indoor dairies. As a successful artist, Ward may have left the dung heap behind, but he had two major problems in class-ridden England: the wrong accent and a lack of the social graces of his younger rival, Edwin Landseer, who also had the right connections: he painted Victoria and Albert in 1842, and the Queen knighted him in 1850.

Still, Landseer did not do cows, so, should they want a memento, the owners of a champion horse or cow had no choice but to have it painted, and Ward was the best artist for the job. Ward also had a lucrative sideline selling prints, but he found himself cheated after 1830 by pirated editions of his famous works that sold all over the country for a couple of shillings each. In 1847, he applied for and was granted a pension from the Royal Academy.

He painted on until 1855 when he was paralysed by a stroke, but he gloried in new technology. In 1852, he travelled from London to Ramsgate in a train with his wife, a journey taking three hours and costing 10 shillings each in first class. He said: 'Surely this may be called the march of Intellect'.

After his death, Ward's 1821 work, *The Bull Family*, was exhibited at the Crystal Palace and admired all over again, before his son sold it to an admiring nation for £1500.

## The rise of the photographer

Inspired by the Exhibition of the Photographic Society in January, *The Times* declared the photographer to be an artist. 'He must be the master of the laws of perspective, light and shade, and colour and composition, no less than those of photographic chymistry'.

Sadly, 1859 was the first year when there were no miniatures in the Royal Academy summer exhibition—photography had replaced them. In April, the Artists' Benevolent Association met in London for their 44th anniversary festival, and those present included Landseer and William Frith. The Chairman noted in a toast that during the previous year, 73 cases of distress were relieved, and that the damage photography was doing to miniature painters meant there would be more calls on their funds.

In France, Charles Baudelaire was savage in his review of 'The Salon of 1859':

> If photography is allowed to supplement art in some of its functions, it will soon have supplanted or corrupted it altogether, thanks to the stupidity of the multitude which is its natural ally. It is time, then, for it to return to its true duty, which is to be the servant of the sciences and arts—but the very humble servant, like printing and shorthand, which have neither created nor supplemented literature. Let it hasten to enrich the tourist's album and restore to his eye the precision which his memory may lack; let it adorn the naturalist's library, and enlarge microscopic animals; let it even provide information to corroborate the astronomer's hypotheses; in short, let it be the secretary and clerk of whoever needs an absolute factual exactitude in his profession ...

Practical uses were being actively sought. An Australian explorer, Herschel Babbage, had already taken photographic equipment into the desert in 1858. Catalogues of lace and embroidery now used photography, said *The Times* in January. The late emperor Nicholas of Russia had been kept abreast of progress on a new bridge at Kiev, by photographs. After a collapse on a railway site, the contractor had taken photographs, which might be expected to be tendered in some future lawsuit. These, said *The Times*, were infallible records, and photographic copies of wills, settlements, conveyances and deeds might also be of some value.

US manufacturers used photography in silverware catalogues. One heavy machine builder used photos to record each manufacture/sale while textile makers used the new medium to generate patterns to be printed on calico. At the end of the year, the British Army's Royal Engineers declared that eight NCOs were to be appointed to teach new and important skills: photography, lithography and printing, telegraphy, surveying, fieldworks and the laboratory. Each would hold the rank of colour-sergeant, with one shilling and sixpence per day extra.

In June, Lt Walker, 79th Highlanders, used photography to copy targets on the rifle range, rather than transcribing the results by hand. With the rise of a volunteer rifle corps, this would be invaluable, said *The Times* and, in July, *Scientific American* agreed. In the same issue, the American journal warned of a photographic reproduction of a bank note so perfect that when the original and copy were shuffled, nobody could distinguish them.

Still, as the journal had reported three weeks earlier, photography of counterfeit bank notes allowed lithographs to be distributed—there were 144 banks in Massachusetts alone, making the detection of counterfeits a challenge. A week earlier still, readers were told there was a device—the stereoscope—that could be used to compare two notes. If they differed even slightly (because one was counterfeit), the variant portions stood out boldly.

The stereoscope, able to superimpose two images for a 3D view was mainly seen as an amusement device, though makers of instruments and stereograms were keen to persuade parents of their educational value. Five advertisements in one issue of *The Times* at the start of 1859 respectively announced or offered: stereoscopic novelties, a stereoscopic fair, stereoscopes, and educational collections of views of Italian cities and towns. The fifth sought a stereoscopic camera with lenses which 'must be quick and equal working'.

Stereograms of the Elgin marbles were offered. Where objects could not be stolen, unlike the Elgin marbles which had been, stereo views were available by year's end showing Egypt and Nubia, Athens, Constantinople, Cairo, Prague, Italy, the Rhine and Switzerland. A catalogue would be mailed post-free by A. Gaudin and Brothers.

In England, a craze that began in 1850—for very tiny photographs—continued. *Scientific American* reported that M. Amadia (probably an error for Amadio, see below) had taken a portrait of Dickens no larger than a pin's point, and another worker had produced the Lord's Prayer of similar size. 'Under a magnifying glass, it appeared as perfect as if printed in bourgeois'. Another advertisement offered a set of 36 stereoscopic pictures of Rome.

This was planned as the first of a series offering both education and entertainment to young and old.

Mr Amadio of Throgmorton Street advertised in *The Times*, offering his photographic portraits of Charles Dickens and Albert Smith, almost invisible to the naked eye, but viewable with a microscope. The Smith referred to was probably Albert Richard Smith, a humorous writer and medical man who had adapted some of Dickens' stories for the stage.

In England, a Mr Mills was charged with destroying two signs, one of which featured a photographic portrait of his wife. The judge ordered him to pay restitution but, arguing that the signs' exhibitor was entitled to no sympathy, he limited the damages to two guineas. A few years later, Alfred, Lord Tennyson would have been sympathetic to Mr Mills, because his friend and neighbour on the Isle of Wight, Julia Cameron, acquired a camera in 1863, and promptly began taking and distributing likenesses of the poet. 'I can't be anonymous by reason of your confounded photographs,' the poet fumed. The Age of the Paparazzi loomed on the horizon.

Millais painted his last Pre-Raphaelite work in 1859, *Vale of Rest*. He needed money to keep Effie (the former Mrs Ruskin) and a growing family, so he turned his back on art and churned out cloying and sentimental works like his 1860 lovers *The Black Brunswicker*, all the way down to his 1885 *A Child's World*, better known as *Bubbles* from its use in Pears Soap advertisements.

By 1859, photographs were being printed on blocks of wood to help the engravers of woodcuts, pictures had been taken at a depth of three fathoms in Weymouth harbour, and Gaspard Felix

Tournachon, also known as Nadar, had taken photographs from a balloon—no mean feat with the heavy equipment of those days. One other aspect of photography emerged early, said Baudelaire:

> The love of pornography, which is no less deep-rooted in the natural heart of man than the love of himself, was not to let slip so fine an opportunity of self-satisfaction ... the world was infatuated with them.

It is unlikely Baudelaire knew of the predilection of an Oxford logician, the Reverend Mr Dodgson, (aka Lewis Carroll), for photographing semi-naked little girls. One of his models was Alice Liddell, Prince Leopold's later brief love interest. Most biographers believe Dodgson meant no harm: the parents knew about the photographs, which were returned to the subjects at puberty, but such behaviour would certainly raise a few eyebrows today.

While the moral attitudes of the Victorians were often a mere charade, perceived foreign indecency incensed the Victorians. Regulations were introduced in the Australian colonies to cover 'obscene and indecent' imports, in line with the British *Obscene Publications Act* of 1857. *Scientific American* thundered that 'Hindoo literature abounds in folly and filth, and much of it is unfit for perusal. It is, however, springing up afresh with the beautiful truths of the Bible infused into it'.

The journal had observed a few months earlier that the books of the Japanese were small and of little value in morals or science, '... while not a few are licentious and obscene'. The voice of the prude was heard in the land.

## TEMPLES OF CULTURE AND KNOWLEDGE

It was always thought good for common people to improve their minds, and so galleries, museums, zoological and botanical gardens, parks and even the occasional playground opened all around the world. The National Gallery of Scotland was founded and the National Portrait Gallery opened at 29 Great George Street, London, in 1859, after having been approved in 1856.

Collections were rationalised in the 1850s. In 1858, the natural history collections of the US Patent Office were transferred to the Smithsonian Institution. The British Museum had opened its doors on 15 January 1759, and a century on, it was packed. In 1858, a trustees' sub-committee discussed moving the plant collections to the botanic gardens at Kew, hearing evidence from assorted Kew botanists, who favoured the move, said *The Times* in March. Professor Owen, who worked on vertebrate fossils at the museum, agreed—but he hoped some fossil plant material would stay at the museum.

Letters were received from Sir Charles Lyell and 'Mr C. Darwin of Bromley Kent', who thought a national collection of botany, like an art collection of paintings, should be housed in London. The sub-committee reported against the proposed move, and on 15 March 1859, the trustees were ready to start considering a proposed increase of accommodation for the Museum collections.

In a single week, visitors to the Museum of Patents, South Kensington, numbered 1093 in the morning sessions and 1433 in the evenings. The Crystal Palace had a half-crown day on

Winter garden design for Alexandra Palace, Muswell Hill, London.

17 March, with 1195 paying entrants and 3149 entering on season tickets. A One Shilling Day in January netted paid admissions of 872 and 483 visitors on season tickets.

Around the world, cities could see the need for parks and gardens. The Missouri Botanical Gardens was established in 1859, as was its Singapore equivalent. In Australia's gold-rich Ballarat, the first Buninyong Municipal Council reserved an area of 10 acres for botanical gardens in the city. Australia was particularly well off for botanic gardens, with established gardens in Sydney, Hobart and Brisbane, all well before 1859.

Walled cities, said *Scientific American*, always had extensive fields and parks within the walls. The vibrant city of New York had lacked these, exciting the derision of foreigners, but at the end of 1859, the 842 acres of Central Park would change that, with eight-and-a-half miles of carriage road, five miles of bridle road and 30 miles of footwalks. Work had started in June 1858, and would be completed by about fall, 1860. Boston's Public Garden was established in 1859 on 108 acres of filled land owned by the Commonwealth of Massachusetts.

Vienna had a zoo in 1752, Paris opened one in 1794, London in 1828, Berlin in 1844. Now came the catch up, as zoos opened in both Philadelphia and Copenhagen in 1859, while others were contemplated. A zoo opened in Melbourne in 1860, and another in Central Park, after work started on it in 1859. The Museum of Comparative Zoology at Harvard opened in 1859 under the direction of its founder, Louis Agassiz, who had made his name by proposing the controversial idea of ice ages. He garnered public money for a grand scheme to 'demonstrate God's master plan for the animals, emphasizing the individual creation of every single species'. The Massachusetts legislature granted $100,000 to the Museum of Comparative Zoology which Agassiz had founded at Cambridge, said *Scientific American* in April. The money was to be paid in instalments, raised from the sale of state lands, but only on condition that private subscriptions had exceeded that amount.

By the end of the year, Agassiz' overtly creationist scheme came under attack, when Darwin's book came out. Agassiz, as a professional zoologist, may not have already known about Darwin's and Wallace's brief Linnean Society papers on natural selection in 1858, but he certainly would have known that most scientists had long accepted that species changed. Still, he fought the idea, and continued to do so, even up to his death in 1873. By that time, his former students and even his own son had accepted the Darwinian model of evolution by natural selection.

If the intellectual edifice of Louis Agassiz was left in ruins, his building remains as a pleasing edifice. For the most part, the Victorians built to last.

## ARCHITECTURE

An upper middle class Victorian family in Britain might have a 10-room house with six bedrooms and a bathroom with a piped water supply. There would be a library, a drawing room, a dining room and a kitchen in the cellar, where one of the servants would sleep. No data are available for 1859, but in 1851, a labourer's home in rural England had an average of 4.4 people in a thatched cottage, one bedroom with four or more people in a square 10 feet by 10 feet. Later in the century, another bedroom or two was often added, along with a scullery or kitchen. The walls were commonly single-course stone, so the interior was generally damp.

The Gothic Revival style was thought most appropriate for libraries and institutions of learning, but in a private house, even when the exterior was different, a home's library was quite likely to be got up in faux Gothic Revival. Other larger 1859 buildings in this style include Government House, Perth, Western Australia and the University of Sydney. Britain had the entrance to the Assize Courts in Manchester, though the high point probably came in 1862, with the Albert Memorial in London, commemorating the newly deceased consort of Queen Victoria.

Canada's capital buildings were designed after a competition, organised in 1859. Gothic Revival was selected for the parliament building, two adjacent administrative buildings and the Governor-General's residence. The style was chosen because it best represented parliamentary democracy: it was, after all, the style used for the Houses of Parliament at Westminster. And because it was fashionable.

As ever in human history, fashion remained an imperative in work, in science, in death, and even more so in the lighter moments of life.

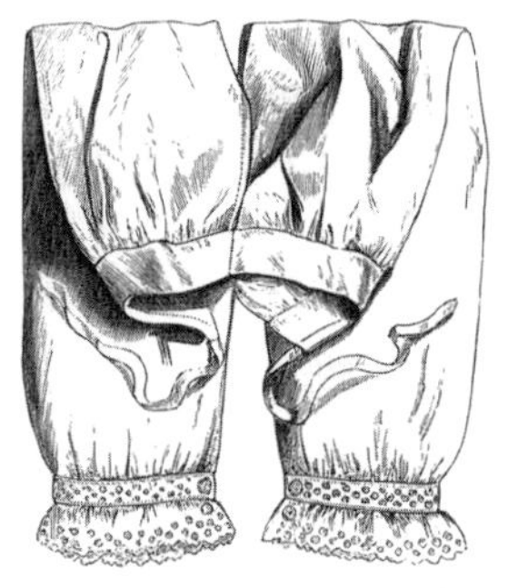
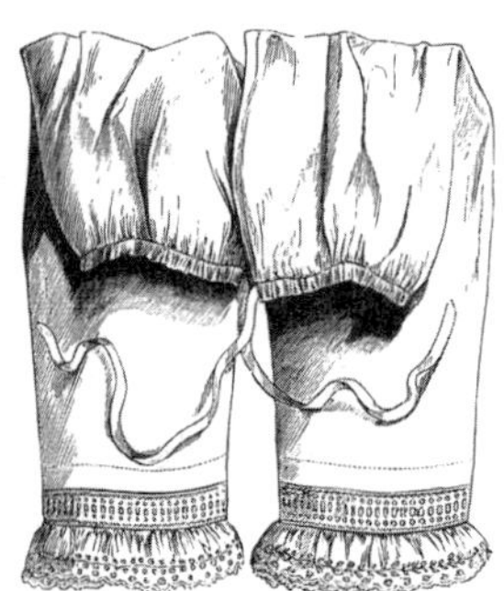

# 7:
# A life of leisure

*Fashions, after all, are only induced epidemics.*

George Bernard Shaw, *The Doctor's Dilemma.*

The young Charles Darwin had no beard, old Charles Darwin was bearded. British men were generally clean shaven until soldiers returned from the Crimea with beards, though 'literary men' had beards sooner. In early 1861, Abraham Lincoln explained, just before his inauguration, why he would be the first US President to have a beard in office: an 11-year-old girl, Grace Bedell, wrote and suggested he should grow one because his face was thin; but if fashion did not sway Lincoln, it must surely have influenced Grace Bedell.

Men who dyed their grey beards or hair used lead-based dyes and many of the colours they used contained a variety of heavy metals. Women who dyed their hair were also at risk, but they had more to fear from arsenical dyes in their gowns, while everybody was threatened by proximity to the then-fashionable green wallpaper, dyed with arsenic compounds. It was fashion.

Fashion was just as lethal to the whales which supplied whalebone for corsets and crinolines, and it had been as bad for the beavers which had provided the fur needed to make gentlemen's hats until the 1850s. Then the varnished silk hat took over, but *Scientific American* did not like them, saying the hard-shell hats were a menace. Some hats had gauze tops for ventilation, but most did not. While felt hats are somewhat porous and so somewhat ventilated, silk plush hats were saturated with lac-varnish and completely airless. They needed perforations at or near the band, argued the reporter. Later in the year, William Warburton obtained a patent for a machine that used heated points to perforate the sides of a hat, a system that *Scientific American* recommended for any headwear coated with varnish.

Ladies' underwear was also causing some worry. With the development of the crinoline, where hoops of whalebone, wire or other stiffening converted the dress into a giant bell, exposure of the limbs was more likely. Legend has it that ladies, fearful of being blown on their sides by wind or swooning, suddenly wanted more modest underwear, but the evidence is, at best, scanty—unlike the new underwear, it seems.

## HIDDEN FASHIONS

The problem with writing social/domestic history is that all too often, people at the time did not record the details of their everyday activities. We cannot always get the details we need to understand everyday life, even two life spans in the past. William Perkin's mauve, discovered in 1856, gave ladies a new (and safe) fashionable colour. It was new and different enough to draw attention, so we know it came into production in a major way in 1859, even if it was derived from the noxious remnants of coal gas and coal oil production.

Sometimes an industrious journalist filled in the background while earning his fee, as when Septimus Priesse wrote about making and colouring bonnets. He described mordanting straw bonnets with an ounce of iron sulfate in two gallons of water, boiling them for an hour, then hanging them out to dry, adding that chip or leghorn straw needed less mordanting than other types of straw. Next, the bonnets were boiled in two gallons of clean water for an hour with half a pound of broken nutgalls and half a pound of logwood, two common dye sources of the time.

The next step, he said, was to leave two ounces of best glue in two quarts of water overnight before boiling to dissolve it, and straining the glue, now referred to as size. Then, readers should soak the bonnets in the size, one at a time, before removing them, sponging off the excess size and drying before carefully shaping the hat, or placing it on a block to dry. The result would be a nice black bonnet. A few details of daily life might be deemed too intimate and, so, inappropriate but more often, they just seemed too ordinary to be spelled out, so we sometimes have to rely on inference, or unpublished sources.

Diaries and letters are useful records of everyday activities. Because Eliza Edwards' letters were describing life in Hawaii to her family in New York, she included ordinary matters like donning rubber boots to walk through knee-deep rushing water. In her diary, Caroline Cowles Richards, a young girl in upstate New York, revealed how a friend pierced her ears for her so she could wear earrings—as well as revealing the fashion influences she experienced:

> Mary Wheeler came over and pierced my ears to-day, so I can wear my new earrings that Uncle Edward sent me. She pinched my ear until it was numb and then pulled a needle through, threaded with silk. Anna would not stay in the room. She wants her's done but does not dare ... It is nice, though, to dress in style and look like other people. I have a Garibaldi waist and a Zouave jacket and a balmoral skirt.

Not everybody agreed with fashion. Empress Eugénie of France had pioneered the crinoline, but she declared in 1859 that she was giving it up. Unmoved by the edict of a mere empress, the style held on. It was claimed in the press that in Istanbul, the Ottoman sultan had, by decree, imposed a limit upon the luxury of the Turkish women of high position, and ordered certain changes in their costume. This does not ring true: perhaps it was put about by somebody annoyed by the challenge of trying to pass crinolined ladies on a narrow street, or to fit them into a pew, a doorway, an omnibus, or a carriage.

On the other hand, the crinoline was good for business. It had sparked 100 patents in France in four years: four in 1855, 16 in 1856, 30 in 1857, 37 in 1858, and 13 by July 1859. Covered steel crinoline (wire) sold at 50 cents a pound, and about three-quarters of a pound was needed for one hooped skirt. The estimated usage in 1859 was five million pounds. At the end of November, *Scientific American* reported that in Derby, 950,000 hoop skirts had been made since 1 April, using 9,100,000 yards of tape and 445 tons of steel.

A Mr Wappenstein in Manchester received a patent in 1859 for making artificial whalebone from animal horn. This would be cut in long helical strips which were then flattened and heated before being coloured. They were suitable for use in both umbrellas and crinolines.

According to cricket lore, round-arm bowling was developed by a cricket player's sister who found that her crinoline got in the way of conventional underarm bowling when she helped him practise. She is usually named as Christine or Christina Willes, and is supposed to have come up with her innovation in the early 1800s, half a century before the crinoline. Her dress may not have been embroidered, but it appears that the story was.

Round-arm bowling was possibly more lively than an underarm delivery, but it was still no great threat to the batsman. An advertisement in the *Victorian Cricketer's Guide* of 1859–60 offers batting gloves, wicket-keeping gloves, and 'leg guards' but no protective boxes for the groins of the male players. There was no real call for them at the time, as the umpire would call 'no ball' if any bowler raised his arm above his shoulder.

## SPORTING FASHIONS

To most Americans and Canadians today, cricket is a mystery, but Abraham Lincoln attended a cricket match between Chicago and Milwaukee in 1859, and a professional All England cricket team toured Canada and the US during the year, playing five matches, the first overseas tour in any sport. Taking a cricket team to the US back then was not as bizarre as it sounds from today's perspective.

In 1859, cricket was very popular in the mid-Atlantic states, in Boston and the New England factory towns, but it could also be seen in Baltimore, Savannah, New Orleans, Cleveland, Cincinnati, and even San Francisco; there were perhaps 300 or 400 clubs.

Cricket even inspired American inventions. In March, M. Doherty of Boston patented a cricket bat that would not jar or bruise, but which would send the ball further. The blade had a wooden shell filled with cork or similar material, while the handle was hollow and contained a strip of whalebone.

In 1859, however, the baseball craze started to bite, and soon cricket would be eclipsed. The writing was on the wall for cricket in September, when the ball game for the Massachusetts state championship caused enough interest for several railroads to issue excursion tickets to Boston's Agricultural Fair Grounds.

Many new sports arose around 1859, perhaps because the lawnmower was now mature technology and delivered smooth playing fields for all sorts of game. The original mower was developed in 1830 from a machine used to remove the nap from

Female archers, Crystal Palace, London.
Archery offered perspiration-free exercise.

cloth, and it allowed smooth, true grass surfaces, something almost impossible to create with a scythe or with grazing animals, but organic mowers still had a presence. The *Illustrated London News* reported in the middle of 1859 that a lightning bolt had struck a sheep in London's Hyde Park, summarily terminating its earlier sterling grass control services. After about 1860, horse drawn and then motorised mechanical mowers did most of the work.

Lawn tennis was developed in 1859 by a solicitor, Major Thomas Henry Gem and his friend, Perera, a Spanish merchant. The two lived in Birmingham, England, and played a game on Perera's croquet lawn that they termed 'pelota', based on a Spanish ball game. Fifteen years later they formed the Leamington Tennis Club, which laid out the rules of this game that became known as tennis.

In 1868, the All England Croquet Club was created to control croquet and to unify the laws. They leased four acres at Wimbledon in 1869, and tennis courts were added later when the croquet fad waned. The club changed its name to the All England Lawn Tennis and Croquet Club in 1899, and has held that name to the present day, even if the world just thinks of it as 'Wimbledon'. Croquet had become a British craze in the 1850s, and the first recorded croquet game in the US was at Nahant, Massachusetts in 1859.

Football was also emerging. In May, the rules for 'Australian Rules' football were developed, though the Football Association, the founding body for the world game (or 'soccer', if you must) only wrote out its rules in 1863, with Rugby codes developing about 1870. In 1859, Allan Robertson, the world's first golf professional died, still hating the new-fangled 'gutties', the golf balls with gutta percha in their hearts. They made the game too easy, he thought.

Not all sports owe their birth to lawnmowers. Polo (literally 'horse hockey') was started in India in 1859 by the Maharajah of Manipur, Sir Chandrakirti Singh. It was the year lacrosse was named as Canada's national sport and the first ice hockey game appears to have been played in Halifax in 1859 (ice hockey became Canada's national winter game in 1994).

The first modern Olympic Games were staged in Athens, not in 1896 but in 1859! A Hellenic grain merchant named Evangelos Zappas convinced the Bavarian-born King Otto I of Greece to patronise an Olympic festival at Athens. Otto was driven out of Greece in 1862, which caused the second Olympiad to be somewhat delayed, and these days, we take the second attempt of 1896 as the first of the modern series.

A Meyerbeer opera, *Le Prophète,* opened in 1849. It featured apparent ice skaters (roller skaters), but that and an 1849 'spin-off' ballet, *Plaisirs de l'Hiver ou Les Patineurs* helped to make roller skates popular, while *Les Patineurs* remains in the orchestral repertoire today. In 1859, the Woodward skate with vulcanised rubber wheels, was unveiled in London, but people did more than demonstrate their strength and agility. They went to see the experts in action.

## CIRCUSES, ACROBATS AND STRONG MEN

The Romans had wanted their bread and circuses, the folk of 1859 would settle for just a circus, but it was rather less barbaric than the Roman namesake. The 'equestrian circus' began in London in 1786, but 1859 was the year that the flying trapeze was added to the bill. The world's first flying trapeze circus act was performed on 12 November at the Cirque Napoléon in Paris by Jules Léotard, 21, who had practised at his father's gymnasium in Toulouse. He wore the daring (for that period) tights which still carry his name.

A whole series of daring young men followed him, but few were as daring or showy as Charles Blondin, the tightrope walker. Starting on 30 June, Blondin made 21 crossings during the summer on a rope 1100 feet long stretched 170 feet above the boiling waters of the Niagara Falls, from Prospect Park on the US side to the Canadian side.

On 17 August he carried his manager across the gorge on his back. The trip lasted 42 minutes and included 42 rest stops. *Scientific American* was scathing: 'We did not suppose that two such fools existed on this hemisphere. The idea of such a thing is enough to congeal the blood'.

Doctor George Winship, a 25-year-old physician who trained in Cambridge Massachusetts could raise himself by either little finger until he was half a foot above it. He could also raise 200 pounds by either little finger and lift 926 pounds dead weight, without the aid of straps or belts, said *The Times*.

Closer to home, *Scientific American* used the same figures a week earlier, suggesting that both journals drew from the same original source or press release, as there was no time for the American material to have crossed the Atlantic.

The American account says Winship was due to give a lecture in Boston, but fainted twice. He attributed this to the atmosphere being close and impure, though others thought it was because he had not spoken in public before. His lecture was on physical education, but the aptly named Boston *Atlas* reported that the strong man proved an infant. Winship seems to have disappeared from public notice thereafter.

For his own pleasure and the amusement of others, a gentleman in Liskeard, Cornwall fashioned himself a suit made solely from 670 rat skins, collected over three and a half years. It included neckerchief, coat, waistcoat, trousers, tippet, gaiters, shoes and even a rat hat.

It was a measure of the way people were being urbanised that dogs were now seen more as companion animals than as work assistants. The world's first dog show was held at Newcastle-upon-Tyne in June, while Birmingham held another show in November. To this day, Britain's National Dog Show is organised by the 'Birmingham Dog Show Society (founded 1859)'. The Battersea Dogs' Home was established in 1860.

There was a poultry and pigeon show at London's Crystal Palace in January. No doubt a few scientists who knew Darwin's ideas would have dropped in to view the displays, because the artificially selected breeds of birds were central to Darwin's arguments about what could be achieved by selection of another sort, natural selection.

Perhaps they took in a theatrical show while they were in town, but perhaps they did not, because many still thought the theatre lacked propriety.

## THEATRE AND PUBLIC MORALS

Increasing levels of education and literacy, combined with evening schools and greater leisure as workers began to win on the 'hours' front, meant people had more time to think, and more time for leisure. The theatre was widely seen as a place of loose morals and easy virtue, but audiences still flocked to the theatres. In New York, Dion Boucicault opened *The Octoroon; or, Life in Louisiana* on 5 December. It was seen by many as an attack on slavery, though others saw it as defending slavery.

Boucicault's most lasting effect came when he suffered piracy of his work in the US in 1853. With R.M. Bird and G.H. Boker, he got a copyright law through the US Congress in 1856, but it took many more years to get clear and enforceable legislation. *The Octoroon* included a slave auction scene, an exploding riverboat, and also an up-to-the-minute plot device when photography was used to solve a crime.

Dionysius Lardner Boucicault was born in Dublin, and may have been the illegitimate son of Dionysius Lardner, a famous 19th century science writer, a man with an eye for the ladies and very close to the family.

Still Boucicault kept the name of his mother's Huguenot husband, 26 years her senior, even after she moved to London with Lardner and her son. Then young Dion got the acting bug, and helped change the way theatre was seen in Britain and the US. He even toured Australia in the mid-1880s, outraging the Australian

Even popular flea circuses arrived in London from Europe.

middle classes by marrying an actress in his company who was 44 years his junior. This did a great job of encouraging the curious to come and see the scandalous pair perform, but it helped to confirm the view some still had of theatrical types.

In Indianapolis, the manager of the Metropolitan Theatre offered to hold a benefit for the local Widows and Orphans Asylum. There was soul searching, with pragmatic board members eager to accept the donation. Others drew the line at accepting 'tainted money' from theatre folk, and in the end, the offer was declined.

Still, many upright citizens wanted entertainment. Asked by a reader for recommendations of the safe and innocent family amusements in New York, *Scientific American* suggested 'Drayton's Parlor Opera' at Hope Chapel, Broadway. The performances were remarkably spirited, very amusing and 'perfectly free from the usual evils of theaters'. All parts in the 'entirely unobjectionable' performance were played by Mr Drayton and his wife. The journal described a collection of paintings known as 'Waugh's Italy', as 'also one of the harmless exhibitions which are well worth seeing'.

In London, Benjamin Webster's brand new Adelphi Theatre offered private boxes with a saloon holding six for two guineas, family boxes holding four for a pound, stalls two feet wide were five shillings to three shillings, while pit stalls with elbows and cushions were two shillings. The cheapest seats were sixpence, but for this, you would see a sketch, *Mr. Webster's Company is Requested at a Photographic Soiree*, followed by the comic drama *Good For Nothing*, and a grand Christmas pantomime *Mother Red Cap*. Like most of the other pantos, it offered an extravagant 'Transformation Scene', and had named stars playing Harlequin, Columbine, Clown and Pantaloon.

Taking *The Times* around his theatre on 22 December 1858, Mr Webster showed how theatres had improved. All refreshments would be under his control, so there would be no extortionate prices. Spacious cloakrooms for the ladies were on offer, with all the requisites for the toilette and no fees would be charged for caring for cloaks or bonnets. The staff would all be women, reducing extortion or fee taking, and the whole theatre from pit to ceiling, was fireproof. The many exits would allow the entire audience to leave almost instantly. The reporter was ecstatic, writing 'No transformation which this year's Adelphi pantomime can furnish will be half as great or half so striking as that which the audience will behold in comparing the old theatre with the new.'

The Victoria Theatre may also have been fireproof, but just five days later, people were killed in a fire scare that began when a boy in one of the boxes set fire to a box of matches. There was a puff of smoke, some women screamed fire, and panic set in as the people in the gallery burst out, opening the doors. There were two performances scheduled for 27 December, and the house could hold 3000 people, a third of them in the gallery. This was reached by a spacious staircase with four landings, with a ticket box on the third level. That made the first three flights effectively a vestibule, closed off by a door below the fourth flight.

At the early performance, 800 people were in the gallery, and a crowd was blocking the stairs up as far as the closed door, waiting to get in for the next show. When the fire panic began inside and doors at the top of the stairs opened, the new arrivals surged forward even as others struggled to get out. After 15 minutes, sixteen people were dead.

## THE HALLS

Until they closed in 1859, Vauxhall Gardens, founded in 1661, was a popular haunt of well bred young men of a certain kind. The Gardens offered food, drink, pantomimes, fireworks, dancing, balloon ascents—and dalliance, which was why the young men went and nice young ladies didn't, but the fashion died. Once the gentry stopped coming, the gardens closed, and many patrons moved to the music halls, a new craze which spread well beyond London.

Wilton's Music Hall opened in 1859, with splendid décor of white, gold and mirrors, a 'sun burner' chandelier with 300 gas jets to light the stage, and room for 1500 patrons. Bought by the Methodist Church in 1884, it became a soup kitchen, and saw its proudest hour as the HQ for those who saw off Mosley's bully boys in the 1936 Battle of Cable Street. Wilton's was the location where 'Champagne Charlie' was first heard (and some say the first English can-can took place there). The suggestiveness of acts generally grew over time.

Charles Morton let women into his Canterbury Hall from 1852, but created two more halls in 1859 and 1861, while Glasgow's Britannia Hall was started in 1857 and opened in 1859. After 1859, the numbers swelled as more people acquired enough money to afford tickets. London had 200 halls by 1868, 347 in the 1870s.

In Massachusetts, Theodore Parker preached each Sunday in the Boston Music Hall, and in London, Charles Spurgeon made use of the Surrey Music Hall. To some of the very pure, all things are impure, and one of Spurgeon's deacons urged against using 'that devil's house', but the preacher was more practical: 'We did not go to the music-hall because we thought it was a good thing to worship in a building usually devoted to amusement, but because we had no other place to go'.

All the same, Spurgeon used the financial clout of a regular Sunday morning booking to get the management to agree to close the hall on Sunday nights. When the foundation stone of a new tabernacle was laid in July, *The Times* attacked the proprietors of the Surrey Music Hall for wanting too much for the use of the hall for a celebratory breakfast, pointing out that the hall was getting £780 a year for rentals, and was likely to do so for some time to come.

In the end, the Surrey Music Hall management decided the loss of Sunday night trade was more than they could wear, and Sunday night concerts began, late in the year. On 14 December, *The Times* reported that Spurgeon had decided not to preach again at the Surrey Music Hall, as it was now opened on Sunday evenings for music, 'although chiefly sacred'. He moved his services to the rather smaller Exeter Hall instead, until the new tabernacle opened in March, 1861.

The 1843 *Theatres Act* forbade legitimate drama in British music halls, though dramatic interludes and sketches were allowed. Christy's Minstrels were on tour in Ireland in January, but the Ohio Minstrels (15 vocalists, dancers and musicians) were in London, as were 'The Coloured Opera Troupe', who, dressed in court costume, offered 'refined NEGRO CONCERTS' at the Oxford Gallery, before a provincial tour in February.

In January, *The Times* indicated that Mr Dickens was offering a few more 'Christmas readings', with performances on 6 and 13 January, for prices ranging from one shilling to four shillings. Elsewhere, Mr Barnum repeated his lecture on 'The Art of Making Money' with prices from one to three shillings.

For those wishing to avoid theatres, there were glee clubs and home entertainment. The same page of *The Times* revealed that evening parties could be entertained by 'Her Majesty's Ventriloquist and Magician', Mr Wellington Young, who had entertained HM one night in 1846, but was still trading on it. A Young Married Lady was willing to entertain on the pianoforte or play for juvenile balls and evening parties for 3s. 6d. an evening. More upmarket accompaniment could be had from quadrille bands, though budget quadrilles could be danced to just a cornet and piano.

## MUSICAL PERFORMANCE

In Manchester, 'The Halle Orchestra' had been created in 1857, and some said it was quite as good as that of the London Philharmonic Society. In an age before modern entertainment media, music held a central place, and in an age when all music had to be 'live', many more people took part; not always a good thing.

The tastes of the masses applied. Mendelssohn's *Wedding March* had been arranged by a Devon organist in 1847, though it only became fashionable after the wedding of the Princess Royal in 1858. Then there was *La Prière d'une Vierge,* known to the English as

*A Maiden's Prayer.* This piano solo was let loose in Paris in 1858, and introduced into England in 1859. Percy Scholes said of Thekla Badarzewska, its composer who died aged 23:

> In this brief lifetime she accomplished, perhaps, more than any composer who ever lived, for she provided the piano of absolutely every tasteless sentimental person in the so-called civilized world with a piece of music which that person, however unaccomplished in a dull technical sense, could play.

Muscular musical criticism was all the go in 19th century Britain. Bach had died in 1750, Beethoven in 1827 and Louis Spohr in 1859. In *The Mikado,* Gilbert offered a list of punishments which included being forced to listen to 'Bach, interwoven/With Spohr and Beethoven/At classical Monday Pops'. The Popular Concerts, or Monday and Saturday Pops, began in 1858, and offered mainly chamber music. At the time, Bach was regarded as suitable fare only for the strong-willed, strong-boned, teeth-gritted musicologist, and Beethoven and Spohr were somewhat out of fashion as well.

Beethoven was still heard, if rarely. The Bradford music festival in August was attended by the Queen, her Consort, the Prince of Wales, seven earls, a duke, an archbishop, a bishop, nine assorted mayors and lord mayors as well as other dignitaries. Handel, Mozart, Rossini, Donizetti, Meyerbeer, Beethoven, Verdi, Mendelssohn, Weber, Bellini and Cimerosa were all played, as well as Léopold de Mayer and Hullah.

Mostly names we still know, but no Brahms, no Bach, no Vivaldi, no Haydn. The first large-scale performances of Bach's entire Mass in B minor took place in 1859 in Leipzig, but that would have been too Catholic for Bradford.

The morning program featured *The Messiah,* the evening had Mozart's *Jupiter Symphony* and a new cantata called *The Year* by W. Jackson. This was 'William Jackson of Masham', a Bradford organist who was held in high esteem, but who has now faded into oblivion as Bach had then. Let us hope that in a hundred years, Spohr and Jackson will be back in the catalogues and playlists. Léopold de Mayer (or de Meyer), a virtuoso pianist, seems to have been largely lost from view, as has John Pyke Hullah who combined with Charles Dickens in a now-forgotten comic opera called *The Village Coquettes,* which ran from 1836 to 1837.

Hullah appeared in the controversy over standard pitch that began around 1859. In a time before cathode ray oscilloscopes (CROs), the A above middle C could, and did, vary from 420 Hz to as high as 457.2 Hz (on New York Steinway pianos—in London, Steinways used A=454.7), though A=440 was more common. The high figure can be determined from a tuning fork, still in existence; some of the lower figures come from looking at the tensions that can be withstood by early keyboard instruments. (A side issue: how could scientists or musicians measure frequencies before the CRO? Answer: they mainly relied on a variety of mechanical stroboscopes, which returned highly accurate measures for standard tuning forks.)

*Scientific American* reported that a meeting in London had decided a uniform pitch would be desirable. The French C above middle C was 522 vibrations a second, Hullah used 512, others used a lower tone. Jenny Lind (famed as 'the Swedish Nightingale') argued that the high pitch then in vogue was harming singers' voices. Still, 1859 was a good year for opera.

Verdi's *A Masked Ball* was produced, Gounod's *Faust* and Offenbach's *Orpheus in the Underworld* and *Genevieve de Brabant* were all premiered in Paris. Wagner completed *Tristan und Isolde.* In Madrid, what was described at the time as a 'hired crowd' (presumably in the pay of a rival) booed Madame Grisi until she fainted during a performance of *Norma.* Berlioz revived Gluck's *Orfeo ed Euridice* the same year.

The hymn *Nearer My God To Thee* became popular, recycling an older tune, but opera became involved with politics. A scheduled Naples performance of *A Masked Ball* had to be moved: it depicted the assassination of a Swedish king and any mention of assassinating royalty was unacceptable to the authorities. An 1859 La Scala performance of Bellini's *Norma* somehow included a reference to the enemy eagles, meaning the Roman eagles.

The Austrian censors had banned this, because the Hapsburg crest also featured an eagle. The phrase 'Viva Verdi' was heard, but VERDI here meant *Vittorio Emanuele Re D'Italia,* Italian for 'Victor Emmanuel, king of the Italians'. The booing of Madame Grisi was nothing in comparison to the cheers of the Italian mob, enflamed by opera and patriotism.

In other areas, Daniel Decatur Emmett, a northerner and the son of abolitionist parents, had his *Dixie* first performed on 4 April in the Mechanics' Hall, New York City. In January, Louis Moreau Gottschalk completed *La Nuit des Tropiques* subtitled *Symphonie Romantique* and composed his *Columbia, caprice américaine* for piano, Op. 34, D. 38 (RO 61) in 1859. There are distinct hints of both

Schubert and ragtime about this piece which features a familiar tune, *My Old Kentucky Home*.

Brahms completed his *Concerto No. 1* for piano and orchestra during the year. It may have been originally devised as a sonata for two pianos, and then as a symphony before becoming a piano concerto. Camille Saint-Saëns gave the world his *Symphony number 2*, but it would be another 27 years before his third or *Organ symphony*. Louise Farrenc was part of a large and well-connected artistic family, but after her daughter died of tuberculosis in 1859, she stopped composing.

Connections were everything, even in music. *The Times* reported that Meyerbeer had been less than happy at the 'thunder' produced for his *Pardon de Ploermel*, but then he heard stones and mortar falling to the ground through a long wooden trough on a building site. This thundered much better, so he hurried to the theatre and had a trough made, but found that stones gave too hard a sound. He decided grape shot would be ideal but the manager told him these were munitions which could only be procured with the permission of the government. Unfazed, Meyerbeer wrote to Marshall Vaillant, Minister of War, who made the army's stores at Vincennes available to the composer, giving him the thunder he desired.

## Musical instruments

Musical instruments were changing. The now-forgotten *harmonicor* was invented by Louis Julien Jaulin, while on 20 December, Henry Steinway Jr took out US patent 26532 which covered the over-stringing of grand pianos. This was a turning point for

Steinway, and their pianos won many awards around the world thanks to this innovation. Ignaz Bösendorfer died during the year, and his son Ludwig took over that family's piano making business.

In early January, Howard Glover announced a chamber concert featuring a number of vocalists and two solo instrumentalists: Miss Emma Green and Mr Henry Blagrove, 'first violon of the Philharmonic'. The 'violon' is just a French 'violin'. Using the word in English was a little pretentious, even then. Henry was another of those concerned with standard pitch, and his brother Richard was an eminent concertina player.

A Wheatstone concertina and harmonium price list of the time shows prices ranging from £1 16 shillings, to 12 guineas for '... a full-compass instrument as used by Signor Regondi and Mr Richard Blagrove'.

If the English public might enjoy the concertina, they loved Handel. Born Georg Friedrich Händel, he lived for most of his adult life in England, and George Frederick Handel became a British subject in 1727. He had died in 1759, but where Bach, Spohr and Beethoven fell from favour after their deaths, Handel stayed in the public's eye, and ear—and in the repertoires.

The first Handel Festival at the Crystal Palace in 1857, was just a warm-up for the centenary. In one issue of *The Times* on 3 January, the public were alerted to the Green-Blagrove concert and also informed of the celebrations of Burns and Handel that would take place during the year. London would be the main centre, but most cities and towns would offer something. In 1859, the British public would be Handel hogs.

## FOOD AND DRINK

They were about to be food hogs as well, going on a later account in *Scientific American* which said customers at the Handel Festival ate 1600 dozen sandwiches, 1200 dozen pork pies, 400 dozen Sydenham pastries, 800 veal and ham pies, 480 hams, 3509 chickens, 120 galantines of lamb, 240 forequarters of lamb, 150 galantines of chicken, 60 raised game pies, 3022 lobster salads, 2325 dishes of salmon mayonnaise, 300 score of lettuce, 41,000 buns at a penny each, 52,000 twopenny buns, 32,249 ices, 2419 dozen 'beverages', 1150 dozen ale and stout, 403 Crystal Palace puddings, 400 jellies, nine tuns of roast and boiled beef, 400 creams, 350 fruit tarts, 3500 quarts of tea, coffee and chocolate and 485 tongues. 'The consumption of wines, which was enormous, had not been ascertained when our account was made up', it concluded.

In some parts of the world, people may have been starving, but London was not the only major city that did very nicely. In 1858, New York city's inhabitants accounted for 191,374 beeves, 10,128 cows, 36,675 veals, 551,479 swine. Each week, a thousand beeves came to New York from Illinois alone. Lake Superior farms exported 7 million tons of corn and oats and more than 3 million bushels of wheat in 1859. Ten years earlier, it had been a mere 1400 bushels of wheat, said Archer B. Hulbert, who had looked into the matter.

Feeding habits changed quickly. At the Great Exhibition of 1851, London had only half a dozen restaurants listed in a guide for visitors, all of them pricey. By 1859, restaurants were becoming common around Soho, where many of them were opened by foreign immigrants. Their leader, if they had one, must surely have been Alexis Soyer, who started out as the second cook to Prince de Polignac at the French Foreign Office. During the July revolution of 1830, he left Paris for London and took the post of chef at the Reform Club, where he cooked for a number of English aristocrats. On the morning of Queen Victoria's coronation on 28 June 1838, he served breakfast for 2000 guests.

Soyer was a Victorian celebrity chef. In 1847, he wrote to the press about the famine in Ireland, and went to Dublin at the government's request to set up kitchens to serve soup and meat at low cost. He also wrote a sixpenny book, *Soyer's Charitable Cookery,* and gave the profits to charity. He resigned his Reform Club position in 1850 to open his Great Exhibition restaurant at Gore House, now the site of the Royal Albert Hall, but took a loss of £7000 by the time he closed, three months later. He spent the next four years promoting his various books and also his 'magic stove', a spirit burner which could be used at the table.

In 1855, he wrote to *The Times,* proposing to go to the Crimea at his own expense to advise on feeding an army. While Florence Nightingale changed the way the sick were treated at Scutari, Soyer changed the way soldiers were fed, beginning with the hospital diet sheets. For the healthy soldiers, he designed an ingenious field stove which the British army only stopped using recently.

When he died in 1858, times had changed. If Soyer had opened a restaurant then, he probably would have had more luck than he did in 1851. He was famous enough after his death for his name to be evoked across the Atlantic by *Scientific American* in an engineering context:

> Soyer always maintained that there could be no good cooking where the scales, the watch and the thermometer were not in constant reference. These instruments are as essential to steam-engineering as to cookery.

Isabella Beeton was another celebrity of her age. Mrs Beeton, as we recall her today, died of puerperal fever in 1865, having made herself famous with her books on household management, which appeared in print when she was barely 22. Her recipes and rulings on management appeared first in her husband's magazine, *The English Woman's Domestic Magazine*, which ran for the three years from 1859 to 1861, then her book *Household Management* followed in 1861. Like the magazine, it marks an era when women who could read still had to manage their households: literacy was filtering down the social ladder. Contrary to folklore, she offered no recipe for cooking rabbit beginning 'First, catch your rabbit ...', but the format she adopted for her recipes is still used today.

In fairness, Mrs Beeton was not the first, just the most successful, of her kind. Elizabeth Ellet produced *The Practical Housekeeper: A Cyclopaedia of Domestic Economy* in 1857 (a man of his times, her husband, William Ellet spent his final years as a chemical consultant for the Manhattan Gas Company until he died in 1859). Eliza Acton wrote a number of recipe books, including *Modern Cookery for Private Families* in 1845 and *The English Bread Book* in 1857 before she also died in 1859, the year in which Mrs. M.H. Cornelius published her *The Young Housekeeper's Friend* in Boston.

'Poverty and oysters always seem to go together,' said Sam Weller, '... the poorer a place is, the greater call there seems to be for oysters.' By the 1850s, natural oyster beds were almost depleted in many places. Notwithstanding Sam's view, a meal of fried Olympia oysters and eggs was usually the most expensive on the menu in California, making it the meal lucky miners would order when they struck gold. The dish gained the nickname 'Hangtown Fry' after a condemned man asked for the dish as his last meal—or so the legend runs.

There was a new condiment for the daring to try: in 1859, a Colonel White made his first batch of hot sauce from 'Tobasco' chillies and offered bottles of it for sale. A slightly different formulation was patented in 1870 as 'Tabasco', benefiting those who think wasabi tastes bland on its own.

The demand for oysters, with or without sauce, was such that the oyster beds near Ceduna in South Australia, an area only settled and exploited after 1836, were already under threat from over-harvesting. Around the Bassin d'Arcachon on the coast of France, southwest of Bordeaux, a place where wild oysters had been taken since Roman times, the locals were forced to start farming oysters in 1859.

At the end of the year, a plan was announced to use a diving bell to harvest oysters from the bottom of Long Island Sound, and it was suggested that parties would be able to go down in the bell, collect their own oysters, and consume them at a depth of six fathoms. *Scientific American* offered an explanation for not eating oysters when there was no R in the month. English oysters, said the reporter, spawn for about six weeks, starting around June!

'Refrigerators' were on sale in 1859 but these were just ice boxes, used to make drinks cooler. Francis Bacon had died in 1626 of a chill, said to have been triggered by experimenting with a chicken stuffed with snow to see if it would keep longer, but refrigerators were rarely used to stop food going off. All the same, ice was a useful commodity in the mid-1850s. Ice ships loaded up in Massachusetts and rounded Cape Horn to make sales in Australia.

James Pimm was the landlord of an oyster bar in London's financial district. He sold his gin-based Pimm's No 1 Cup from the 1840s, and with backing from some of his customers, he began bottling and selling it in 1859. Beer and ale were still preferred as safer than water, with coffee and tea, even later in the century when water supplies improved. Water was feared as tainted and impure, and in most cases, rightly so. Even supposedly safe alcoholic drinks carried risks, and *Scientific American* listed tests for adulterants.

Copper in beer could be detected by evaporating the beer down to 'the consistency of an extract' and then burning it, treating the ash and looking for a blue trace that became darker when ammonia was added. Lead in beer could be detected by adding sodium sulfate and looking for a white precipitate, and the article went on to explain how a large variety of other adulterants could also be detected by chemical tests on beer extracts.

The water was full of germs and no good, the food was full of adulterants and no good, the air was full of poisons and noxious: it was enough to make you sick. Then again, in 1859, there were ever so many ways to make people sick.

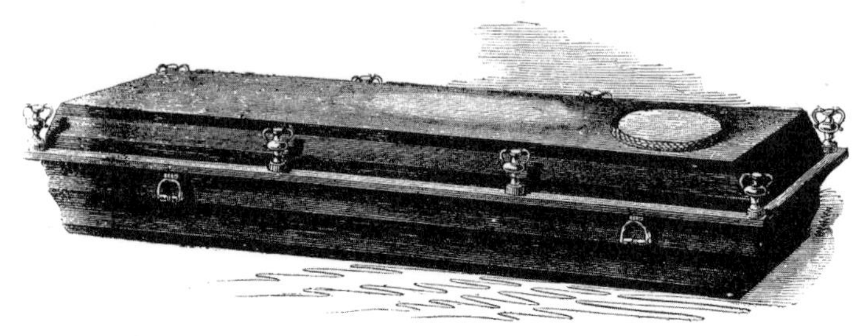

# 8:
# The ills of society

*With fingers weary and worn,*
*With eyelids heavy and red,*
*A woman sat, in unwomanly rags,*
*Plying her needle and thread –*
*Stitch! stitch! stitch!*
*In poverty, hunger, and dirt,*
*And still with a voice of dolorous pitch*
*She sang the 'Song of the Shirt.'*

Thomas Hood, *The Song of the Shirt*, 1843.

On New Year's Eve, 31 December 1858, an unnamed woman starved to death on the streets of New York with her child beside her. As 1859 dawned, 'Charity' wrote to the *New York Times*, asking why people were not more vigilant. The police should spend more time peering into nooks and crannies, the writer declared. The *Baltimore Patriot* reported the death by starvation of an eight-year-old boy. A post mortem revealed that his stomach was shrivelled, and only two small pieces of apple were in it. In Ireland, the *Portadown Weekly News* was exultant, later in the year:

> We have now passed the time when, in former years, accounts of the appearance of blight in the potato crop usually began to appear. As yet, fortunately, there have not been even rumours of disease in the growing crop unless from one or two Western districts; and the potatoes, which are in very abundant supply at the Irish markets, are excellent quality. Owing, however, to the severe frosts at the commencement of May, a considerable portion are small in size, and we fear that this complaint is rather general.

So there would be no starvation in Ireland in 1859, but severe drought and famine hit Kansas across 1859–60, affecting some more than others. In March, men were dying of starvation as they headed for Pike's Peak, and the way up Smoky Hill became known as the 'Starvation Trail'. Goldfields must have often been places of famine: there is a Starvation Flat in California, and a Starvation Creek in Victoria, both in gold-bearing areas.

New Yorkers were shocked to read the sad tale, telegraphed from St Louis, of the Blue brothers. Daniel, Alexander and Charles

Blue, had set out, with a man called Soley, to win their fortunes at Pike's Peak. Starving, the men died one by one: first Soley, then Alexander, and then Charles, and each time, the survivor(s) ate parts of their freshly dead companion, in order to survive. Daniel was one of some 500 miners who returned to St Joseph, reporting terrible conditions, and those accounts, highlighted by Daniel's harrowing tale, were enough to slow the rush, but only briefly. The hunger for gold overcame any fear of food shortages.

Starvation was worst for those travelling in challenging circumstances. John McDouall Stuart was trying to find a way across central Australia for the projected Overland Telegraph in 1859, and he explained in his journal that he found conditions trying:

> As my rations are now drawing to a close [for we started with provisions only for three months, and have been out now for three months and more], I must sound a retreat to get another supply at Chambers Creek. It was my intention to have sent two men down for them, but I am sorry to say that I have lost confidence in all except Kekwick. I cannot trust them to be sent far, nor dare I leave them with our equipment and horses while Kekwick and I go for the provisions.

## POVERTY, STARVATION AND SURVIVAL

Henry Mayhew published his *London Labour and the London Poor* in 1861, but the data he used had been collected in the early 1850s. He told his readers that London costermongers, street sellers of fruit and vegetables, might break their fast with a cup of coffee and two slices of bread and butter, then lunch on 'block ornaments', pieces of cheap meat, sold at twopence a half pound and cooked in the tap room, or if they were in a hurry, pies, either fruit or meat, but never eel pies, because 'they know that dead eels go into them'. The costermongers could always buy more food if they had money, but travellers and explorers were often far from shops or inns.

Starvation is central to Darwin's explanation of evolution, but starvation as such was not a major killer of humans. Malnutrition and associated diseases usually carried the young and the weak off before they could starve to death from a lack of calories. Somehow, most people found enough food to keep going, just.

Mayhew said London's streets had a thousand instrumentalists playing barrel organs, tomtoms, violins, hurdy-gurdies, bagpipes and Irish pipes. There were 250 balladeers and 50 'Ethiopian serenaders', most of them whites in blackface. An 1864 campaign led to the *Metropolitan Police Act* which gave police powers to prosecute the musicians: supporters included Thomas Carlyle, Charles Babbage and Alfred, Lord Tennyson. Babbage in particular, was known to chase organ grinders down the street when they played near his home.

There were 3000 brothels and 9000 prostitutes known to police in London in 1859, though a figure of 80,000 prostitutes in London is often quoted. Babbage did not, as far as we know, chase after them, though their numbers suggest many other men did, even in times when 'Victorian morality' was at its height.

New York had a Home for the Friendless between Fourth and Madison Avenues at 32 East 30th Street. This useful, philanthropic institution was founded in 1824, by private munificence and it was under the care of the American Female Guardian Society. Its object was to afford '... a place and means of protection for destitute, respectable females without employment, friends or home, and within the age and circumstances of temptation: also for friendless children of both sexes, until they can be committed permanently to the guardianship of foster-parents or worthy families, who will train them to respectability and usefulness'.

The names of these institutions seem to have been chosen to inflict a body blow. Sydney offered the Randwick Destitute Children's Asylum from 1858 to 1915. This held 846 children at its peak, the offspring of parents who had either abandoned them or been judged unfit to care for their children.

The boys tended a vegetable garden, milked cows, made shoes, raised silkworms, and did carpentry, baking and cooking. The girls helped with the cooking, cleaned rooms, and did the laundry and needlework.

Breakfast was bread, tea and porridge, dinner was meat and vegetables, supper was bread and tea. Between 1863 and 1891 when statistics were kept, 174 children were buried in a small cemetery there.

At least the shoe-making boys need not have feared competition from machines, but British cordwainers (as footwear makers were called) were on strike in 1859 over the new machinery. *Scientific American* was surprised that they had done so. Prosperity would come from the machines, said the writer, pointing to the way squalid conditions and poverty lead to crime. 'This is a question of social science which deserves more attention than it has yet received by all classes of the community'.

## CRIME AND PUNISHMENT

In London, they said, people were painting birds to look like something rare and foreign, and a 'vulgar rat' was once converted into an elegant microscopic dog as a lady's pet—until its claws grew back, and it ran up the curtains to the ceiling. This tale is as genuine as the recent case where 4000 Japanese were said to have been fleeced by cunning Australians (or New Zealanders) who had sculpted sheep to make them look like poodles. Supposedly, none of the victims thought the feet odd, or wondered that the 'dogs' spurned meat and bleated. We must dismiss this as a shaggy sheep story, just as we doubt the traditional ballad of Mary's canary:

> Sister Mary bought a canary, from the butcher's boy;
> She hung it by her in the dairy, where it was her joy.
> But the bird would never whistle, and she wondered why,
> Until she saw the sparrow's feathers, coming through the dye.

People thought there was a lot of dishonesty about, and that sufficed. Most 19th century convicts were victims of poverty or upbringing, and much education was aimed at breaking the cycle. The skills of rat-doctoring and sparrow-dyeing were not in the curriculum, but perhaps the techniques on offer were less remunerative.

In 1823, William Cobbett said a third of all English prisoners were in gaol because of the Game laws. Under the *Night Poaching Act* of 1817, anybody hunting armed at night could be sentenced to transportation for seven years. After 1831, when people could be licensed to hunt game, British landowners could still retain the rights to hunting on their tenants' farms, meaning farmers could not legally kill rabbits and hares on their land.

When 306 convicts landed in Western Australia in 1862 on board the transport *Lincelles*, 112 had been convicted in 1859. They included two army deserters, five soldiers who had shot at, struck or offered violence to superior officers, seven arsonists, eleven rapists, twelve who had wounded others, seven who had assaulted others, one who had attempted to poison, four who had committed manslaughter and one murderer, five who had counterfeited, coined, uttered or forged, two who had received stolen goods while the rest had committed burglary, theft, larceny, housebreaking, stealing or warehouse breaking. There were no poachers, though two of the cargo had been found guilty of poaching in 1860. One man had 'returned from transportation'.

By 1859, the established Australian colonies had said a firm 'no' to transportation, but a few convicts were still being transported to less-developed Western Australia. When Dickens' *Great Expectations* was published in 1861, there were no hulks left for a convict like Magwitch to escape from, but his informal departure from the hulks was set in the distant past, and Magwitch's later crime of returning from transportation was still a punishable offence. Get out and stay out was the message the courts gave their villains.

At the start of the year, subscriptions were being solicited to establish a London Reformatory for Adult Male Criminals: £1000 was needed, £300 had been raised. The *Reformatory Schools Act* of 1859 set up procedures for dealing with offenders under the age of 21, but gaol did little to improve prisoners or to discourage new offences. Australia at least put the convicts in a healthier climate where there was a demand for unskilled labour, as well as the chance to gain skills.

Three London bankers, Sir John Dean Paul, William Strahan, and Robert Makin Bates, were tried at the Old Bailey in 1855 on charges of fraud after their bank suffered losses in mining and railway drainage. They were found guilty and sentenced to be transported for 14 years.

Their names appeared in the Cavan *Observer* in 1857 as members of a party of 400 convicts about to sail for Australia on 25 August. With them would sail 'Robson, the Crystal Palace forger; Redpath, who committed the forgeries on the Great Northern Railway Company; Agar, the railway guard, who committed the great gold robbery on the South-Eastern Railway' and the 'notorious bank forger, Barrister Saward, alias Jem the Penman, the putter up of all the great robberies in the metropolis, for the last 20 years'.

For some reason, the bankers never left Britain, and four years later, they were quietly released. In 1860, J. Ewing Ritchie, writing about gaol conditions, explained how prisoners who cannot stand the hard work of building breakwaters like Sir John Dean Paul, '... are employed in mending clothes, in making shoes, in baking, and brewing, in the school-room, and other offices necessary in such an enormous establishment ...'. In short, that which was part of daily life for boys in Sydney's Randwick Destitute Children's Asylum was punishment for England's wayward bankers.

In June, three convicts engaged in building a river wall near Chatham dockyard seized a boat and tried to obtain an unauthorised early release, but they were caught. There were usually a thousand convicts on this work and escape attempts were rare, said *The Times*. Gentlemen prisoners like Paul and his friends stayed put and got official early releases.

Being of the right class even made it easier to escape punishment altogether. While serving as a congressman in 1859, Daniel Sickles shot and killed Philip Barton Key, son of Francis Scott Key, the author of *The Star-Spangled Banner.* The younger Key had been engaged in a love affair with Teresa Sickles, and the shooting took place on a city sidewalk in Washington DC in front of numerous passers-by and within view of the White House. Sickles was acquitted on the grounds of temporary insanity. Officially returned to sanity, he served as a Union general in the US Civil War, and lost a leg to a cannonball at Gettysburg.

In the run-up to the Civil War, David S. Terry, an Associate Justice of the California Supreme Court killed US Senator David C. Broderick in a duel in California. This was one of a number of attempts by the pro-slaving lobby, the 'chivalrists', to cow their opposition, but a California jury acquitted Terry of the murder he had clearly committed. In 1889, US Special Deputy Marshal David Neagle gunned Terry down when he attacked another judge.

One crime was uncovered in 1859, for which there appears to have been no punishment. In 1848, John Payne Collier produced an old folio of Shakespeare which had many apparently ancient changes on it. Based on this, he published a new edition of Shakespeare, and later sold the folio to the Duke of Devonshire, who made it available to the British Museum. Nicholas Hamilton from the Museum's Department of MSS, wrote to *The Times* about the folio on 22 June. After a delay, the letter appeared on 2 July.

Prisoners exercise: above working the treadwheel, below merely walking.

In the letter, Hamilton noted that the binding dated from long after Shakespeare's period, probably from about the time of George II, and that there was reason to believe the amendments were made after the volume was bound. The amendments in pen were the right style for the assumed age (close to Shakespeare's time), they shadowed annotations in pencil which were in a more modern hand, marks which had been inadequately rubbed out. It was a total fake. Nobody was ever able to prove that Collier was to blame, but he was implicated in other forgeries, so most people regarded him as the culprit. His only punishment was to be forgotten.

At other times, the accuser was forgotten instead. John Bunyan was accused in *Scientific American* of taking the plan of *Pilgrim's Progress* from an earlier French work. In the garbled report, the name of the alleged original author was wrong (it was de Guileville), and the name of the accuser was wrong (the daughter of Nathaniel Hill had edited her father's work, but was misidentified as Catherine Isabella Curt, rather than Cust). She had published her claim in a collection of her father's notes in 1858, but it seems since to have been largely forgotten, based as it was on some rather vague similarities. The following week, however, *Scientific American* cited the story in protesting that one of their own stories had appeared over somebody else's name in a Chicago journal. Trampling the rights of living authors alarmed them more.

Burglars would have been alarmed by a new burglar alarm for use in hotel rooms. It was a three-barrelled pistol, designed to be hung so it would drop if anybody opened the door. This would set off three shots, frightening the burglar and waking the occupant. It could also be loaded with ball or shot and carried in the waistcoat pocket: struck firmly on the base, it would fire three shots at once. Said *Scientific American*: 'no traveler should be without one'. How many sojourners, absent-mindedly opening their doors, might have been shot by this device, their cases being dismissed with a verdict of suicide?

## SUICIDE, GALLANTRY AND EXPLODING COFFINS

Suicide in Britain had long been a serious offence. Until 1823, suicides were buried in a crossroads with a stake through their heart; until 1832, they had to be buried between 9 pm and midnight—by then they could be buried in the churchyard, but no service could be said. Until 1870, a suicide's personal property was forfeit to the crown. In 1567, Edward de Vere, the 17th Earl of Oxford, killed an unarmed under-cook by the name of Thomas Brincknell. Brought to trial, he claimed to have been practising fencing when the victim ran upon the point of his sword. This was accepted, Brincknell was condemned as a suicide and his widow and child were stripped of their possessions.

The Earl is worth three addenda. First, he is believed by a few to have written Shakespeare's plays; second, Raphael Holinshed was on the jury that let him off: Holinshed's chronicles were later used by the author of Shakespeare's plays; third, the Earl was familiar with foreign places, giving him the background to write the plays. What is less well known is the explanation John Aubrey offered for his travels:

> The Earle of Oxford, making of his low obeisance to Queen Elizabeth, happened to let a Fart, at which he was so abashed and ashamed that he went to Travell, 7 yeares. On his returne the Queen welcomed him home, and sayd, My Lord, I had forgott the Fart.

It would be enough to make you want to go out and slit your throat, wouldn't it? He didn't, though.

The only recourse for the family of a suicide was to have the victim found to be insane, in which case his property was not forfeit. John Sadleir, MP and Junior Lord of the Treasury, had engaged in fraud and forgery on a massive scale, so on a night in February 1856, he went to Hampstead Heath with prussic acid and a case of razors. The acid sufficed, so the razors were superfluous, but he left notes expressing remorse at the harm he had done, and on that basis, was found to be sane by a coroner's jury. Dickens modelled Mr Merdle in *Little Dorrit* on Sadleir, while Trollope drew on him for part of Melmotte in *The Way We Live Now*.

A clergyman was found dead in his Taunton hotel room. A letter was found which indicated that he planned to meet his son, so a jury found he had died of an accidental overdose of prussic acid which was medically prescribed. In this case, the benefit of the doubt appears to have been granted appropriately.

Robert FitzRoy's uncle, Viscount Castlereagh, and Captain Pringle Stokes, his predecessor as captain of the *Beagle* had both committed suicide. By 1865, the by-now Admiral FitzRoy was in despair: his *Weather Book* was being attacked as 'prophetic and empirical' rather than being 'accurate and factual'. At the same time,

Charles Darwin had risen to fame. So on 30 April 1865, the Admiral took a razor and slit his own throat. He had shown signs of instability since 1860, when he stood at the Oxford meeting where T.H. Huxley debated with Soapy Sam Wilberforce, the Bishop of Oxford, on the subject of evolution. FitzRoy was present, and stood, holding aloft a very large Bible which he just happened to have with him, asking the crowd to believe God, rather than man.

If suicides were blamed, self sacrifice in war was a different matter. Public adulation went to the (generally posthumous) Victoria Cross winners, to the gallant idiocy that was the charge of the Light Brigade and to Dickens' Sydney Carton, who redeemed himself by doing a far, far better thing. Equally, the brave pioneers who strode off to explore a potentially fatal wilderness were admired. The expectation that a ship's captain would go down with the ship is also at odds with official values on self disposal.

Private George Richardson, VC won his award on 27 April near Cawnpore, in hand-to-hand combat. The *London Gazette* of 11 November 1859 gives a stirring account of his actions:

> Richardson did, despite the fact that his arm was broken by a rifle bullet, and leg smashed by a sabre, rush to the aid of his officer, Lt. Laurie, was attacked by six natives, and that, crippled as he was, succeeded in killing five, and the sixth fled.

In 1859, the future Field Marshal Earl Roberts of Kandahar was awarded the VC for conspicuous bravery in a fight that took place the previous year. Richardson and Roberts were both alive at the start of World War I. Roberts died of pneumonia while visiting

Indian troops in France, Richardson was the Empire's oldest living VC recipient when he rode in a carriage at the head of the 27 August 1921 Warriors' Day Parade in Toronto, Canada.

Fictional suicides, if dressed up with a small amount of self sacrifice, were acceptable. Mrs Gaskell, in *The Manchester Marriage* (1858), described a situation where a man is presumed dead, so his wife remarries. The man returns, manages with the help of a servant to see his 'posthumous' child, and then leaps into the Thames to leave his wife happy. She never learns the truth, though her new husband does. Frank Wilson, the noble suicide, is admired for his self sacrifice.

Real life was harsher. Fanny Coxon of Carlton-le-Moorland near Nottingham was the subject of an inquest in 1858. In her note, she said she had been led astray and was six months gone with child. She was going to a watery grave, and recognised (wrongly) that she could not be buried in the churchyard. She mentioned one young man and indicated that he was not the father. The jury returned a verdict that the death of the deceased was caused by herself while of a sane mind, and the coroner issued a warrant for her interment between the hours of 9 and 12 o'clock.

## Births and deaths

In Britain in the second half of the 19th century, 36 per cent of the population were under 15: this pattern must have held in much of the rest of the world. As mother-of-nine, Queen Victoria was credited with saying that being pregnant was an occupational hazard of being a wife. In 1870, childbirth mortality was estimated in Britain as one in every 204 births. With women commonly bearing 10 children, each woman had about a five per cent chance of dying in childbirth.

Some women died from complications, others died from infections like childbed fever but in 1859, the way the disease spread was identified when Ignaz Phillip Semmelweis finished gathering the data he had been collecting since 1848 to prove beyond doubt that hand washing by doctors prevented the spread of childbed fever, also called puerperal fever, the disease which killed Mrs Beeton. As we will see in the next chapter, the germs that caused most diseases were about to be unmasked.

Tradition has it that 19th century judges were stern and unflinching automatons or avenging furies, unlike the humane and caring British jury, but that was not always the case. A 19-year-old girl, Mary Jones, was found guilty by the jury, of wilful murder. She had been seduced and gave birth, but the child died. Baron Martin was the judge, and he summed up strongly in favour of a conviction on only the lesser charge of concealing a birth, but the jury thought otherwise. Their grim verdict left the judge with no option but to pass sentence of death. He immediately wrote to the Home Secretary, asking that the sentence be commuted, and the Home Secretary agreed.

Disease and death meant funerals for young and old were both common and a major economic drain. Simple funerals cost £4, elegant middle class ones £50, an aristocrat's funeral might cost £1500. The Regent's Park Funeral, Furnishing and Carriage Establishment offered a funeral with a pair of horses, a carriage, a covered coffin and 'every requisite', all for four guineas. Mourning jewellery was made of jet stones and the hair of the deceased, and there were elaborate rules about the mourning attire worn by the women of the family.

Many sorts of innovative coffins were available. Cast iron was so passé, now the fashionable dead in the US were clad in corrugated iron, which was lighter and easier to handle and transport, and featured an airtight window with an India rubber gasket.

*Scientific American* reported that in Britain, terracotta was all the go. They were of the ordinary shape, and the lid fitted in a groove where it was secured by Roman cement, which was an error, said the reporter. The ancients knew better: they did not want their coffins to explode, so they left holes for the gases of decomposition to escape. The same criticism ought to have applied to the corrugated iron coffins, but no deaths were recorded from exploding coffins during the 19th century. There were plenty of other causes of death, though.

Even at the end of the century, in 1899, out of 1000 children born in the better suburbs of Liverpool, 136 of them would die in their first year. In the poorest suburbs, 509 would die in that crucial first year. Smallpox was fast coming under control, but whooping cough caused 40 per cent of under-five deaths, while scarlet fever or scarlatina was the main killer in the five–eight age range. Diphtheria was a big killer of pre-teens but measles and tuberculosis were also major causes of death.

At least the number of desperate murders of unwanted spouses was falling. Divorces did not get you hanged if you were caught, and divorces were now a lot easier to obtain.

## LOVE AND MARRIAGE

Divorce in Britain before 1857 was in three forms. Divorce *a vinculo matrimonii* was a nullification because of insanity, impotence, or a too-close blood relationship. It allowed you to remarry, but made any children of the original marriage illegitimate. Divorce *a mensa et thoro* did not let you remarry, but allowed you to separate. It was mainly for adultery, sodomy or cruelty—which meant violence. In the third variant, some, mostly men, could seek a divorce *a mensa et thoro*, and then sue the spouse for adultery, which meant that Parliament granted you a real divorce that did not make any children illegitimate. Only four of 90 parliamentary divorces granted before 1857 went to women—the cost was well over £1000, which women rarely had.

Forget 'Victorian' morals: in England, one child in 15 was known to be born out of wedlock, but that ignores the hidden cases. Queen Victoria carried and passed on the gene for haemophilia, which does not previously appear in her official family tree. She was probably a new mutation, but it is conceivable (no pun intended) that the queen herself may have been the result of an extramarital alliance of her mother's in 1818.

Lord Palmerston, who became Prime Minister in June 1859, was renowned for his 'flirting'. He was elected to Almack's Club, where he was believed to have 'dallied' with three of the seven patronesses, and even to have tried to seduce a lady-in-waiting at Windsor Castle.

Dickens' wife left him in 1859, perhaps over his liaison with Ellen Ternan, who may or may not have been his mistress, and who may or may not have borne his child, and 1859 was the year Giuseppe Verdi married Giuseppina Strepponi, the soprano with whom he had been living for about 12 years.

In the US, many Civil War dead turned out to have more than one widow because divorce had been effected by simply moving on. There was even a song, popular with Californian miners:

> Oh, what was your name in the States?
> Was it Thompson or Johnson or Bates?
> Did you murder your wife, and flee for your life?
> Say, what was your name in the States?

In 1859, Caroline Drunkel married John Martin, who then went away to war. At war's end, he was mustered out and went to live with a woman to whom he may have been bigamously married in 1863, using a different name. By 1890, the Civil War pension system applied to all invalid veterans, widows and the minor children of deceased soldiers, regardless of the cause of death. A private's widow got $8 a month, officers' widows got more, and there was $2 for a child minor, so the pension was worth fighting for.

Proof of marriage was required with exceptions for Indians and freed slaves, but those categories had to have lived as man and wife and be recognised as such—and there could only be one widow. The files of the time sometimes carry the cryptic mark 'cont. wid.', meaning that a woman was a contesting widow. One soldier, George Bagshaw, had at least four wives recorded.

The method of clearing a marriage was simple for a man: a 'friend' wrote a letter announcing the person's death while off hunting gold or whatever, the man would take a new name and carry on. In an age before photo-ID, credit checks and passports, it was easy to move to a new country and acquire a new name.

An 'Unmarried Gentlewoman of England' published *The Afternoon of Unmarried Life* in 1859, a work with a bleak message. The book set out to help the unmarried woman decide what to do with the rest of her life after she reached 30 without a husband. After the 1851 census, single women were identified as a social problem: part of it would have been war, part industrial accidents,

The Divorce Court, London.

part emigration, at a guess. They were called 'superfluous women', but no doubt some were comforted that deaths in childbirth helped to even out the imbalance. Life was not all beer and skittles in 1859.

Henry Mayhew reported that 10 per cent or less of the costermongering trade were married, while in nearby Clerkenwell parish, 20 per cent were married, because the vicar there married poor couples at Advent and Easter without a fee. No distinctions were made by married and unmarried women, who associated freely with each other as equals. He noted that 'chance children', defined as the '... children unrecognised by any father, are rare among the young women of the costermongers'.

The conservative *Gentleman's Magazine* might huff and puff that divorces were disastrously on the rise, with 288 petitions in the first 15 months, but at the year's end, *The Times* looked back, and said the divorce courts were relieving a lot of suffering. The beneficiaries and rational people could look around and see many greater disasters to worry them.

## FIRE, FLOOD AND OTHER ALARMS

Some two centuries after the Great Fire of London, cities were growing fast, and people feared fire more than ever. Some cared about loss of life, others were more concerned by potential property losses, and a fireproof safe weighing nine tons was made in London for a Spanish bank. This was, sniffed *Scientific American*, another example of an American invention being taken over, but it seems it was more an idea whose time had come.

A trial of a fireproof safe was held in the Royal Botanic Gardens of Sydney in July, with the safe put to an ordeal by fire on reclaimed land at the harbour's edge. A ton of coal, several loads of firewood, tar barrels, packing cases and plenty of loose timber created a structure some three metres square and more than two high that burned most satisfactorily, but the barbecued safe and its contents survived intact.

In 1858, 41 fires were said to be started in London by the careless disposal of unextinguished cigars. In Coulterville, California, the fruit and cigar store of one David Cohen caught fire when rats got in among the non-safety matches in the shop. A Mrs Hauff, said the news reports, started to run out with her two children but somehow ended up in the cellar, which somebody closed soon after, not knowing they were there, and all three smothered. Most of the town was destroyed as the fire spread, said the *Steamer Times*.

New York City's Crystal Palace burned to the ground in October of 1858. Australia reported severe bushfires over the summer of 1858–59, as usual. In August 1859, the old capitol building in Vallejo, California, was destroyed by fire. A fire destroyed a whole block in Houston, Texas, due in part to water shortage, and perhaps in part to squabbling between the three fire companies.

In many parts of the world, fire brigades and fire companies were formed: newspapers and the post made such ideas infectious. Baltimore, Victoria (Canada) and Alamo fire companies were all formed during 1859, and the journals were full of reports of new water pumps that could be hauled to fires and brought into use.

Elsewhere, trains crashed, with 100 injured near Glasgow and seven killed on the Great Western Railway of Canada when an embankment collapsed after heavy rains. A tub fell into a North Shields pit that a waterworks company was drawing water from, dragging one man with it and falling on another. In a coal mine at Ruabon in north Wales, the machinery broke down, leaving miners stranded at the bottom of the shaft. 'It was a curious sight to see the half-bewildered friends of the men feeding them from above by throwing various kinds of food down the pit,' said a report from the scene. The men were later all brought up safely.

Around the world, a number of volcanoes erupted. The Cosigüina volcano went dormant in 1859, but on Java, Mt Bromo produced an explosive eruption in the central vent, and elsewhere in the Dutch East Indies, Raung in east Java and Serua in the Banda Sea were active. Alfred Russel Wallace was near the island of Banda in 1859, and he wrote of the 'ever-smoking volcano', wondering where the heat came from that powered volcanoes.

Wallace was the other naturalist who had stumbled, with Darwin, on the idea of natural selection. Here he is putting his finger on the problem inexperienced people would have with the idea of evolution by natural selection, a process that happens in a timescale that is completely beyond our daily experience, just like the operation of a volcano.

Even extinct volcanoes had a fascination. The Austrian exploration ship *Novara* called in at Auckland, New Zealand in

1858, and Dr Ferdinand von Hochstetter stayed behind to map the isthmus of Auckland with its extinct volcanoes. We will revisit *Novara* in the next two chapters.

The best show of 1859 was Mauna Loa, which erupted, producing a 50 kilometre flow that crossed the Big Island in just 200 days. The flow covered 91 square kilometres with some 383 million cubic metres of lava. Eliza Edwards arrived at Hilo to learn that the volcano had started erupting, and wrote home to New York that it was '… not coming towards Hilo so I entertain no fears on that account'.

By day, red-hot lava looks rather dull and boring, because much of it is covered in a crust of cooling rock. In 2005, I walked six kilometres out onto the slopes of Kilauea to see a lava trickle, trudging over crunchy rock that thrust up, sharp and jagged as broken glass. In the gathering gloom, we began to see small gleams of red above us, where the crusty blocks had moved apart or where lava had oozed between them. Within a few minutes, the hill looked like a torchlight procession, but this was just a dribble of hot rock. In the dark, we snapped on our lights and walked back through the dark to our van. The *New York Times* reported in May that:

> A party of visitors returned to Honolulu after a trip to view Kilauea. They were two days without water, and walking over clinker, had their boots ripped to shreds.

Life in 2005 was easier and boots were better, but Eliza Edwards made a sensible decision. She found she could see the lava flow quite well enough from her husband's ship:

> We were anchored in full view of the Volcano, & it was so clear. It seemed but a little way off, tho the actual distance was about 50 miles—that evening the lava broke out in an entire new place—& rolled down the mountain in two large streams with great rapidity. It was the most beautiful sight I ever beheld for the wind blew strong—& it was as bright as the brightest fire you ever saw. The whole heavens were illuminated by it.

Edwards is an invaluable reporter, because she was writing for a private audience, so could speak her mind. All the same, her audience did not know the area, so she filled in the fine details that would normally be omitted.

> I got some good tea at Lahaina which we made ourselves in the babys nurse lamp—& we thought while we were there we'd send on shore & have plenty of milk if nothing else ... the Capt went on shore as soon as the anchor was dropped to get some—& dont you believe the people are so pious there that they will not allow cows milked on the Sabbath so we couldn't get a drop—now did you ever hear anything so perfectly absurd in your life. I never did—& I just longed to go on shore & tell those missionaries that I thought they were getting beyond Infinate wisdom, for if the Almighty had not designed that cows should be milked on Sunday—he would probably caused them to give no milk on that day—just as he withheld the manna on sunday—& caused a double portion to descend on Saturday.

## RELIGION AND REACTION

Eliza showed a much freer attitude to religion than we assume mid-19th century people had, at least when they were not on the public record. Just as Queen Victoria and many other women were happy to use anaesthetics in childbirth, ignoring the male divines who demanded that they continue to suffer, as specified in Genesis. People were willing to take their religion in small doses, as Charles Spurgeon knew:

> I find in every circle a class, who say, in plain English, 'Well, I am as good a father as is to be found in the parish, I am a good tradesman; I pay twenty shillings in the pound; I am no Sir John Dean Paul; I go to church, or I go to chapel, and that is more than everybody does; I pay my subscriptions—I subscribe to the infirmary; I say my prayers; therefore, I believe I stand as good a chance of heaven as anybody in the world.' I do believe that three out of four of the people of London think something of that sort.

There was certainly a ferment of freedom in the air where religion was concerned. Comrooder Tyabjee, a native of India, and a Muslim, was admitted in London as an attorney in 1859. His oath was sworn on the *Koran*, the new *Oaths Act* having made this possible. The Roman Catholic hierarchy in Britain was restored in 1850, and 1858 saw the first Jew take his seat in the House of

Commons, as the *Oaths Act* now allowed Lionel Rothschild to swear an oath which left out the words 'on the true faith of a Christian'. On the Continent, the Clerical movement was gaining power, bringing a return of conservatism and repression. People still sinned just as much, but nice people didn't talk about it. And the Sabbath was observed, at all costs.

The banks of New York, in the week ending 5 September, exchanged $120,568,395, which *Scientific American* stated as an average of $20,094,629—clearly Saturday was a full trading day, but Sunday was not. On Saturday, 19 November, the *New York Times* said the late Senator David Broderick's life would be celebrated the next day. A procession, organised by Tammany Hall, had been postponed from the previous Sunday, on account of rain. The plan was attacked by some of the clergy, who felt that the sound of 'bands playing funeral dirge and common tune' would disturb innumerable Sunday Schools and congregations.

In Britain, anti-Catholic feeling was easing, at least in some quarters. There had been an annual service of public thanksgiving in the Church of England, inspired by the prevention of Guy Fawkes' gunpowder plot. Ordained by Parliament in 1606, it gave preachers a licensed opportunity to deplore the Church of Rome and all its works but the service was withdrawn from the prayer book in 1859.

Peace was less than universal. The world would soon see biologists and geologists locked in a brawl with primitive zealots over evolution, while most scientific people and religious folk looked the other way, but religion had its own divisions. In the Anglican communion, the low church evangelicals and the high church 'smells and bells' fraternity slugged it out, with many simple folk caught between the warring factions, as we can see in Trollope's *Barchester Chronicles*. In 1861, William Booth broke away from the Wesleyan Methodists to form the Salvation Army.

Then there was the revival movement, of which Charles Spurgeon was a part. It was strong in Wales, but also in Tyneside, and in other places at other times, drawing adherents from other creeds. The costermongers, said Henry Mayhew, had no religion at all, but nobody minded too much, because they held their lack quietly and without trying to convert others.

Charles Bradlaugh wanted to convert people away from religion. He was the first avowed atheist elected to Parliament in 1880, though he was not allowed to take his seat until 1886. As far back as March 1859, Bradlaugh was preaching a gospel of unbelief, and so setting the scene for the frantic attacks of some opponents of science who cast scientists, evolutionists, geologists and atheists into a single basket. Had they the chance, some of them would probably have cast all the evil ones into a single cell, and thrown away the key.

Even today, it is usual for 'creation scientists' to assert that anybody who believes in evolution must necessarily be an atheist, though their language is less extreme than that directed at Bradlaugh. He was depicted as '... toiling, sweating, labouring strenuously to heap slander upon his Creator' and accused of delivering a '... frantic panegyric in honour of hell'. Bradlaugh was barred from speaking at Devonport, but addressed a meeting from a barge, a few feet offshore, where he was beyond the jurisdiction of the town's police.

It was not only atheists who questioned the bible. The year 1860 saw the publication of *Essays and Reviews*, a scholarly look at the Bible which examined some aspects where the Bible is internally inconsistent. Though written by a group of prominent Anglican clergy, it was seen by some as a greater threat to faith than Darwin's evolution ideas.

At a time when some divines demanded a strict literalism in the face of geological evidence, those theologians who urged more cautious interpretations were seen as a bunch of traitors within the camp. In all, the work led to three charges of heresy, while Darwin scored none.

In 1862, John William Colenso, the Bishop of Natal published the first part of *Pentateuch and the Book of Joshua, critically examined*. This reflected his thoughts after a Zulu convert asked if all of the animals could have fitted on the Ark, and whether the Mosaic laws on slavery were inspired by a loving God. Colenso had been a mathematician but he went to Africa after his father's tin mine flooded and his house at Harrow burned down. He found that Numbers 1:45 (603,520 able-bodied Israelites) clashed with another account in Leviticus.

This might have been fine for a mathematician to think, but he said so, which was considered not so fine for a bishop. His archbishop deposed him, but Colenso appealed to the Privy Council and won, though at a cost. Tractarians and Evangelicals all attacked him for involving civil authorities.

People preferred Dr Livingstone, who went on a speaking tour of Britain in 1857, causing a resurgence in missionary activity. He only converted one man in Africa, and was mostly seen as a failure, but he at least regarded Africans as fallen people, rather than as a genetically inferior race.

In Brno, in the Austro-Hungarian empire, a monk called Mendel was just beginning to tease out the ways genetics operated, but ordinary people had a fair idea about how inheritance worked. Then again, they had ideas of sorts about many things—and most of their scientific and medical viewpoints were wildly wrong. In 1859, major aspects of medical thinking were poised to change.

# 9:
# The rise of medicine

*He prayeth best who loveth best*
*All creatures great and small.*
*The Streptococcus is the test*
*I love him least of all.*

Wallace Wilson (attributed).

Louis Pasteur's nine-year-old daughter, Jeanne, died of typhoid fever in 1859, the year her father showed that yeast produces the carbon dioxide which makes bread 'rise'. He deduced that fermentation was a biological process driven by living cells, not a chemical one. He then put nutrient broth in a set of flasks, some sealed, others open to the air, some of them with long curved 'swan necks' which made it unlikely any bacteria could get in.

There was a lot of superstition about life, even in 1859. In particular, people believed in spontaneous generation, where worms erupted in food because they had formed from the food. Johann Baptista van Helmont, for example, had claimed that mice could arise spontaneously in 21 days from dirty clothes mixed with wheat grains, that male and female mice would arise, and be able to breed and produce offspring. So long as people believed that life could just come into being, there was little point in their considering genetics and they were a long way from worrying about how species evolved!

People had known for half a century that tinned meat and other foodstuffs in sealed containers did not go 'off', but what was the key factor? What stopped microbes from appearing in such food? Even Homer, the ancient Greek poet, knew that maggots came from flies laying eggs, but the decay caused by bacteria and fungi was harder to tie down.

The supporters of spontaneous generation thought life needed air, so they argued that sealing the containers stopped life arising spontaneously, and Pasteur set out to challenge this. He must have suspected that small organisms might be in the air and able to get to nutrient broth in an ordinary open flask. His swan-neck flasks let air in, but were designed to make it hard for bacterial spores to get to the nutrient solution: they might drift into the neck, but would settle there, and never reach the broth.

When the solution in his open swan-neck flasks remained pure, Pasteur had torpedoed spontaneous generation, opening the way for people to get serious about evolution. But it was now just a short, if intricate, step from that insight to blaming tiny life forms for causing disease. Tiny life forms had been seen by Anton van Leeuwenhoek in the late 1600s, and around 1720, Daniel Defoe made a reference to these tiny life forms as the causes of disease. The idea was around, but nobody took it seriously.

By 1859, microscopes were getting better, allowing many more tiny life forms to be seen, so Pasteur quite probably ran the test suspecting the truth, and found evidence that confirmed his suspicion. Pasteur may also have read the work of Charles Cagniard-Latour, who died in 1859. Cagniard-Latour had seen yeast budding under the microscope, showing it was alive, and Cagniard-Latour had shown that different yeasts produced different alcohols, but he was by no means alone. In its second issue of 1859, *Scientific American* recommended using a filter with gravel, sand and charcoal, in order to cleanse roof water before it flowed into a cistern.

This would remove 'insects and their ova', washed from the atmosphere, as well as dust, from water. In March, the journal discussed a report about very tiny organisms, so small that one million would not exceed a grain of sand. The idea was out there, but people hesitated to support the unorthodox idea that germs caused disease. Like penguins on an ice floe, nobody wanted to be first.

A certain M. Pouchet in France doubted germs very noisily. After examining dust, he said, he had only twice in a thousand observations found 'the large ova of infusoria in the atmospheric dust'. Rather, he said, 'the corpuscles of which we have heard so much are granules of starch and silica'. Pouchet claimed that smells caused illness, poisonous smells coming from drains and the like. It remained the orthodox medical position, even if the old ideas were slowly failing to provide answers.

A correspondent identified only as 'Fellow of the Epidemiological Society' wrote to *The Times* to say there is a distinction between infection and contagion, and that diphtheria was a disease of infection, caused by 'an unhealthy house or district'. If such a disease appeared, the rest of the family should be immediately removed to healthy situations. Contagion, as the name suggested, was something that spread by close contact. The snag was that almost any evidence for germs could equally be evidence for some form of 'poison', some smell, some gas, some miasma that was passed on by contact and made people sicken and die. Even Semmelweis' hand-washing protocol could be accepted and explained without germs.

## THE EMERGENCE OF THE GERM THEORY

All around the world today, you can see *Eucalyptus* trees, taken from Australia to fight infectious disease. From Inverness to Israel, Kefalonia to Cambodia, San Francisco to Cyprus, Plymouth to the Pontine marshes, the Australian gum tree has been planted in unhealthy spots. The river redgum *Eucalyptus camaldulensis*, the same tree used to fire the boilers of the Murray River paddle steamers, was expected to cure diseases like malaria. If smells caused illness, the tree's sweet-smelling oils should cancel the evil smells that caused infectious disease.

A very thirsty plant, the redgum quickly drains away surface water, stopping mosquitoes from breeding. This encouraged a false inference that the oils were the key to a cure, because infection was hard to observe, and prejudice hard to curb.

Thomas Hobbes, two centuries earlier, saw life in a state of nature as 'solitary, poor, nasty, brutish, and short'. Life for most people in the civilised world of 1859 was all of those except solitary. Disease spread so easily in crowded cities that chains of infection became invisible webs. In 1849, Britain had a major cholera epidemic which killed 16,000 in London alone. Then it spread beyond the capital where the paths of infection were more easily spotted.

Near Bristol, William Budd isolated 'a fungus', a thing carried in water—which he considered to be the cause of cholera. There was a lot of debate in the lay and medical press, but the consensus was that Budd was wrong. His observations on typhoid in the Taw Valley emerged in *The Lancet* in 1859, under the title 'On Intestinal Fever', when he argued that contagion, from person to person, was involved. This, he said, was the view of most country doctors.

Until 1859, soldiers were regularly inspected for venereal disease, but this was deemed to be too humiliating for the men and doctors, so it was discontinued. The 1864 *Contagious Diseases Act* provided for the registration and examination of prostitutes in port and garrison towns. It effectively allowed state brothels to operate in garrison towns with inspections of the women, who were detained if infected until they were cured. The act was repealed in 1886, but the military concern about disease led army doctors to advise prevention in the form of condoms, which probably had a longer-term effect on views about birth control and sexuality.

In 1837, *The Lancet* published lectures by William Wallace, who found syphilis to be contagious. He prepared 'lint moistened with the discharge of diseased patches' from inpatients with secondary syphilis and then took patients without syphilis. He scraped a portion of their skin until it bled, and then he covered it with the lint and watched as syphilitic lesions developed. He then treated the disease with mercury compounds but expressed regret that treating patients at this point meant he could not watch any further progression of the disease, presumably meaning its development into tertiary syphilis.

This work, carried out at the Jervis Street Hospital, Dublin, seems to have caused no great concern, but two French researchers found a different reaction in 1859. Like others at the time, Camille Gibert and Joseph Alexandre Auzias-Turenne believed that syphilis was a later stage of gonorrhoea, so they needed to see if the lesions of syphilis would cause gonorrhoea or syphilis. They deliberately inoculated four young men (one a teen, the others in their twenties), with purulent material from the sores of a patient with secondary syphilis. Perhaps the people of Paris were more caring or better informed, but whatever the reason, the new experiments caused an uproar when Gibert and Auzias-Turenne explained their trials. In fairness, Gibert and Auzias-Turenne were not alone, just more visible than most. And whatever you think of their ethics, their work did establish some important facts that others would benefit from.

Joseph Lister earned his medical degree in 1852 and at the end of 1859, aged 32, he was appointed Regius Professor of Surgery at Glasgow. When he read Pasteur's work he concluded that germs caused putrefaction. The first 'Listerian' operation was to treat a compound fracture, but he later performed a mastectomy on his sister, and Queen Victoria trusted him to remove an abscess from her armpit.

Surgeons were still somewhat looked down on by the public (and by physicians), but they were improving their skills. When surgery was first practised, the surgeon relied on amazing speed to slice flesh and saw bones on a patient only slightly dulled by liquor, gagged and tied to the table, but largely conscious. Anaesthetics were introduced in the 1840s and allowed surgeons to work more slowly and carefully, reducing shock, but infection remained a risk.

Many who died after an amputation arrived with dirt from grubby bandages and dirt-strewn streets already in the wound. It wasn't always a surgeon's dirty hands or methods that infected patients, so the improvement in survival rates was less than instant, even when surgeons lifted their game. Between 1859 and 1870, hospital mortality after major operations was 45.13 per cent. Between 1886 and 1890, it dropped to 7.1 per cent.

There were still two main problems that could run through whole wards: pyaemia, caused by *Staphylococcus* and 'hospital infection' caused by *Streptococcus*. Treatments in the days before antibiotics included Condy's fluid (manganic and permanganic acids with sodium chloride), chlorine, cold boiled water and bread poultices. Mortality could be high, inspiring Canadian Wallace

Wilson to write the verse against *Streptococcus* which appears at the head of this chapter.

Over the next few years, it would become clear that those who tried to stop disease by careful hygiene were succeeding, not because they kept away the bad smells by cleaning, but because cleaning banished the bacteria that caused illness. It would take until the end of the century to explain insect-borne diseases like malaria and yellow fever, but now researchers had a model, a hint, and a fighting chance—so long as they cleaned up their act.

## HYGIENE IN A FILTHY AGE

*Scientific American* reported in mid-1859 that Jews had a greater life expectancy at birth (46.5 years) than Germans (26.7 years), Croats (20.2 years) and Austrians (27.5 years). A scientist called Gatters put this down to race, but the journal thought superior hygiene was a more likely cause. It was an overdue idea, though one may wonder what Mr Darwin might have made of Gatters' idea of survival of the cleanest, since the key factor here was the transmitted culture, not the genes themselves. The human trick of passing on new behaviours is called cultural evolution. It lies behind the way fashion has shaped humanity. Ideas travel far faster than genes.

Between 1848 and 1864, the sewers of London were commonly flushed to get rid of 'cholera-causing' smells, washing the *Vibrio cholerae* bacteria into the Thames, and into the drinking water taken from the river. The big hurdle for the germ theory was that the sanitarians succeeded in making changes that improved survival (though for a different reason), and this got in the way of a full acceptance of the germ theory.

A little earlier, the Chelsea Water Company had its Thames water intake close to the outlet into the Thames of the Ranelagh sewer. London might have become a ghost city, but for the enthusiasm of Londoners for beer—which involved boiling the water during production—and tea, which needed boiled water. In 1873, Australian explorer Ernest Giles offered an example of how far people were willing to go for drinking water:

> I found the hole was choked up with rotten leaves, dead animals, birds, and all imaginable sorts of filth. On poking a stick down into it, seething bubbles aerated through the putrid mass ... the bucket could not be dipped into it, nor could I reach the frightful fluid at all without hanging my head down, with my legs stretched across the mouth of it, while I baled the foetid mixture into the bucket with one of my boots ...

Giles was crossing a desert of course, but some urban household supplies were little better—where they existed. An 1859 Parliamentary enquiry in Sydney was told of an area just down the hill from Parliament House itself: 'Of the houses built recently in Woolloomooloo, not more than one out of a hundred has a drain or a sink'.

About one house in eight had a piped supply, the rest relied on cartage from a pipe in Hyde Park that drained through a tunnel from nearby swamps. The Sydney milk was little better, according to *Crouch's epitome of news, and miscellaneous gleaner* in December, which reported that a member of a parliamentary sub-committee had, on several occasions, seen milkmen watering-down their milk from 'that filthy pond, situated at the junction of Newtown and Parramatta Roads'. There had been 124 deaths in Sydney in November, 75 of them children under five years of age.

Britain's Parliament paid attention after the 'Great Stink' of July 1858, caused by rotting matter and sewage in the Thames. MPs feared for their own safety and passed a bill allowing the Metropolitan Board of Works to borrow money, so work on new sewers could begin. Once again, smell was the target, and the aim was to move sewage from where people could smell it, and so be infected: wrong paradigm, but the right solution! In London, the task of building sewers was entrusted to Joseph Bazalgette, who was chosen as Chief Engineer of the Metropolitan Board of Works in 1855, sponsored by Brunel, Robert Stephenson and Thomas Cubitt. Bazalgette began work in 1859, as soon as the finance was in place. He finished his huge task in 1875.

Londoners must occasionally have been put off by the content offered at their breakfast table by *The Times*. In 1858, readers were told that Mr Lawes and Professor Way had undertaken a minute study of sewage. The dry solid matter expelled by a Londoner weighed two to two-and-a-quarter ounces a day, made up with fluids to some 40 ounces, so London daily produced 152.6 tons of dry matter or 2993.6 tons of moist matter, but added to this was the street run off.

Professor Way showed that each gallon of street water contained 252.6 grains of dry matter, although it was more than three times this when granite roads were involved, while some wood pavements were far lower in solids. In all, the streets supplied an estimated 233 tons of solid matter including trade refuse, diluted by 84,750,000 gallons of water. The precision would have left the reader in no doubt that this was Science.

The question of smell causing disease was under challenge—and being defended. With the best arrangements for disposal, said Dr Letheby, the Medical Officer of Health, there were still dangerous gases produced in the sewers, and this needed action. An engineer who remained unconvinced by the medical men had asked what might happen if the smells were later found to be harmless. The cost of prevention was only fivepence a year for each of the inhabitants, said Letheby, so prevention made sense.

Having set the scene, Dr Letheby explained that smells had indeed been shown to be dangerous, while stinks were lethal. Dr Barker had tested cesspool gases on dogs which responded with vomiting, diarrhoea, loathing of food and general emaciation. 'In other cases the effects were those of a mild continued fever such as is often seen in the dirty and ill-ventilated houses of the poor.' Letheby said there was also a risk of a build up of light carburetted hydrogen or marsh gas (methane), but it was rare.

Indentifying the cause was hard. A Signor Moscati had condensed the 'miasm of pestilential rice fields' in Tuscany, a Monsieur Rigaud had done the same with the marshes of Languedoc, and M. Boussingault tried the same with some of the worst districts of Paris, but their efforts failed. A crowded cholera hospital, the fetid dormitories of household troops and the equally unwholesome dens of London's poor had all been turned over in a hunt for:

> ... the occult [hidden] molecules which are endowed with so much virulence. All that we know of them is that they will reduce the salts of silver and gold and blacken sulphuric acid, and that they contain the elements of organic matter in a state of active change; that they quickly putrefy, and that they form a nidus for the growth of the lowest forms of animated beings.

Doctor Letheby reported in mid-July that a Professor Bourbée of Paris thought the Thames' fetid state could be fixed. The infected bed of the river was to be covered with hydraulic lime (a form of lime used to make mortar) and pebbles. This would form a concrete that would provide a solid channel for flow, even as the causticity of the lime destroyed the putrid organisms, explained Letheby. He warned that the organic impurities in the Thames were now four times as great as a month earlier, and mortality was increasing.

There were still those who felt drinking water was involved in disease. By a dramatic experiment, John Snow (the practitioner who chloroformed the Queen) had implicated a single pump in Soho with an 1854 cholera outbreak. He broke the chain of infection by having the handle of a pump in Broad Street removed. Even earlier, others suspected the water. In 1849, Charles Kingsley (with many theories about the cause of cholera) helped distribute clean water to Southwark during a cholera outbreak. He wrote to his wife:

> I was yesterday ... over the cholera district of Bermondsey. And, oh God! what I saw! people having no water to drink—hundreds of them—but the water of the common sewer which stagnates full of ... dead fish, cats and dogs, under their windows.

There was another side to the clean water campaign: if people had clean, sweet water to drink, they might give up drinking beer. This parallels Count Rumford's invention of the coffee percolator in the late 1700s. Rumford thought the workers of Bavaria drank too much beer, and believed they would be more productive if they drank more properly brewed coffee! Still, whatever the facts, humans were able to reap a benefit.

Samuel Gurney MP (a nephew of social reformer Elizabeth Fry) and a barrister called Edward Thomas Wakefield founded the Metropolitan Free Drinking Fountain Association in 1859 with the aim of giving the public free clean water to drink. It became the Metropolitan Drinking Fountain and Cattle Trough Association in 1867, after it expanded into animal welfare, but humans still had priority. Gurney paid for the first of their fountains, at the corner of St Sepulchre's Churchyard, and it was opened by a daughter of the Archbishop of Canterbury, underlining the moral value of these fountains. The fountain bore the message 'replace the cup', suggesting a low hygiene standard, as we would see it now.

In the west of England, Liverpool was well ahead: at the start of 1859, there were 43 drinking fountains in the city and it was estimated that a thousand people drank daily at each. Later in the year, a drinking fountain of white marble was proposed for Fleet Street. Sir James Duke would pay for it, and it would stand outside St Dunstan's. Cleanliness was clearly to stand next to godliness.

New York had pure water from the Croton and on the Fourth of July, many of the hydrants along Broadway were tapped, with a tin cup attached by a chain (a parallel with the drinking fountain's common cup). This arrangement was designed to let those gathering to see the military parade slake their thirst on wholesome water 'instead of being forced into grog-shops and bar-rooms to obtain a cooling beverage of doubtful liquor'. There should be more of this, said *Scientific American*. 'Pure water, pure air and whitewash are wonderful reformers, and we wish that the value of all three was more highly estimated by our city authorities than they are.' Clean cups, like germs, were still in the future.

The Croton reservoir on Fifth Avenue was where the New York City Public Library is now, between 40th and 42nd streets, then the fringes of the city. In September, the water had a bad smell and taste, but nobody became ill at first. The water commissioners thought there must be a dead horse or other animal there to make plant material rot, but *Scientific American* had other ideas. Drain

London's first drinking fountain, opened 21 April 1859.

the reservoirs and clear out the accumulation of organic material on the bottom, it said.

Clean water was more than a simple matter of health and hygiene. When the Croton supply was interrupted in October, some newspaper offices paid a dollar a barrel for water to run their steam engines, which needed clean water for the boilers. In the age of steam, even if the people could be fobbed off with noisome sewage, the machines could not. Near the end of the year, Queen Victoria opened a new hygienic and economical water supply to Glasgow. The city's old supply was pumped under steam power from the river; now it came by gravity from Loch Katrine, a pristine lake. There were 13 miles of tunnels (70 in all), nine-and-a quarter miles of aqueducts and three-and-three-quarter miles of giant iron tubing. All of the tunnels went through very hard rock, so sometimes a new drill bit was needed after each inch of boring. Tunnels were driven from a number of shafts, but some of them only yielded three yards of progress in a month.

In Ireland, the *Portadown Weekly News* carried a notice from James Carlile, plumber and gasfitter of Lurgan, who advertised that he had changed his premises, and was now opposite Mr Armstrong's pork store, where he would be '... constantly supplied with Force, Farm Yard, Liquid Manure, and Public Pumps; Common Lift Pumps, with Brass, Lead, and Copper Chambers; Water Closet Fittings; Hot, Cold, and Shower Baths; Garden Engines; Bidets and Wash-hand Basins, fitted up on the most improved principles'. The world was getting cleaner.

## WORSE THAN THE DISEASE ...

With no clear evidence of what caused most diseases, cures were generally based on the principle of fighting poisons with poisons. Lister's early antisepsis was based on using a poison, carbolic acid in the form of a spray: it was crude, but it worked, unlike some of the other 'cures'. Diphtheria was sometimes treated by taking a clean clay tobacco pipe, putting a live coal in it, then placing common tar on the coal and smoking it, inhaling and breathing the smoke back through the nostrils.

*Scientific American* reported that a gardener had blackened the inside of his greenhouse timbers with coal tar, hoping to make them absorb more heat, and saw that the spiders and insects all disappeared.

No discussion was offered about the possible effects on gardeners using the greenhouse, or people eating the food that came from the greenhouse. Perhaps it might have prevented death by diphtheria, but then again, so might decapitation.

Scarlatina and diphtheria hit England in 1859, and in Lisbon, the young and newly married Queen of Portugal died of diphtheria after an illness of five days. The first few days of January saw a protracted discussion of diphtheria in *The Times*. The disease was attributed by 'M.D.' to drains not being in good order.

If the water closets in the basement were not properly trapped, poison from the sewers would permeate the whole house, especially the nurseries in the upper stories. 'Another M.D.' said diphtheria was 'a new-fangled name for an old-fashioned disease, malignant quinsy, which in the days of our grandmothers was successfully treated by emetics and bark'. His treatment: quinine and prussic acid, followed by an emetic. Another writer asserted that muriated tincture of iron and honey should be applied to the affected parts with a camel's hair brush. Powdered charcoal in the room would help. According to 'Medicus', cleanliness, ventilation and nutrition were all important. Those needing parish relief should have a liberal allowance of meat and porter—and this should be extended beyond diphtheria to typhus, scarlatina and other infectious diseases. Causes of tuberculosis (TB) were variously identified as rich food, creativity, excessive emotion, disappointment in love, alcohol, stress, waltzing, tight lacing, steam or hot-water radiators, lifestyle, or anything else the writer did not approve of.

Treatments for TB included cod liver oil, bleeding, blisters, calomel, painting the throat with equal parts of chloroform and olive oil with a large brush, and stimulants. Silver nitrate, applied to the fauces, the cavity at the back of the mouth, was also suggested.

Night sweats were common in consumptive cases, and a recommended homeopathic cure was '... rub the patient, every three or four days, all over with olive oil. By this means the perspiration will be reduced, and the strength of the sufferer kept up'. Dr L. Long of Holyoke wrote to the *Springfield Republican* with a novel cure for consumption. A stout stick, three feet long, was to be suspended by a rope or chain, six to eight inches above the head. The patient was to grasp the stick and swing, gently at first, and more vigorously with time. The effect, said the doctor, was to elevate the ribs and enlarge the chest, expanding the lungs. He had used it over 35 years to treat consumption and lung haemorrhage, but also recommended it as a preventative.

Medical histories usually show the first TB sanatorium opening in Poland in 1859 but *The Times* carried an advertisement soliciting funds for the Bournemouth Sanatorium for Consumption in January 1859. This mentions an annual report, and it appears to have opened in 1855. Italy was an escape from TB for some, so the Trollopes and the Brownings went to Florence (where malaria was rife—you can't win them all!). Cannes, the Isle of Wight and Torquay were all considered good for TB, and high mountain air was well regarded.

## Medical advances

Quackery was clearly common, some of it blatant. Adolphe Didier 'the Somnambule' advised the gentry and nobility in an advertisement in *The Times* that he had established a house in Paris at 68 Boulevart (sic) Beaumarchais for the cure of nervous diseases by magnetism, adding that he could be consulted by letter.

For every quack, there were ten serious scientists making useful progress. They proceeded in careful and systematic ways. Albert Niemann isolated cocaine from coca leaves in 1859, after Friedrich Wöhler asked Dr Carl Scherzer, on board the Austrian research ship, *Novara,* to collect some leaves. Wöhler then passed these to Niemann.

During 1859, Adolph Kolbe developed the Kolbe reaction, which made it possible to synthesise aspirin on a large scale, and Frederick Guthrie inhaled amyl nitrate by accident, finding that it led to face and neck flushing and heart palpitations. By 1867, it was being recommended for angina.

By August 1859, London's main sewer tunnels were taking shape.

Anaesthetics were first widely used in the late 1840s, and by 1853, Queen Victoria used chloroform when she gave birth to Prince Leopold. The anaesthetist was John Snow, and if a few rigid churchmen were outraged, the women of England were delighted by the firm lead given by the titular head of the Church of England. In 1854, the Archbishop of Canterbury's daughter (not the widowed Mrs Wilson who opened the first drinking fountain in 1859, but another daughter) made use of chloroform as well. People would ignore strict religious interpretations when it suited them. If they would ignore religion to relieve or avoid pain, they might be willing to ignore narrow theological doctrines when it came to evolution—so long as the science was good.

Darwin's natural selection idea explained a great deal about disease, and that alone would get it a good hearing in 1859 as medicine moved from the realm of folklore to the realm of evidence-based science. It wasn't a total change, as it took time for evolutionary insights to affect medicine, but the confidence of the new breed of scientific medical men provided Britain with a cadre of supporters for Darwinian evolution.

The desperate need, according to reformers like Thomas Wakley was for better medical education that would pass the new ideas and methods around. The University of London awarded medical degrees from 1838 and St Bartholomew's hospital set up a residential college for medical students in 1842. By 1858, all of the London hospitals which took students required them to study both medicine and surgery.

The *Medical Act* of 1858 established a General Medical Council which was authorised to certify degrees and diplomas allowing the holder to be placed on the general register of practitioners. Some 19 licensing bodies offered examinations, though unlicensed practitioners could work without any penalty, unless they claimed a qualification which they did not actually hold. After 1858, the aim of medical education was to produce safe GPs who could be considered gentlemen.

Strange ideas still lurked in the gaps. In 1859, Jonathan Hutchinson was already an assistant surgeon at the London Hospital, but in 1906, he published *Leprosy and Fish Eating* which argued that leprosy ('Hansen's disease') was caused by eating decaying fish. Death and success in preventing it offered the clearest pointers and arguments to medical men. For the moment, medical women had to wait, but their time was coming.

In 1859 Elizabeth Blackwell had her name placed on the register of qualified physicians in Britain after she trained in America, at Geneva College, New York. Sophia Louisa Jex-Blake tutored mathematics at Queen's College for Women in London from 1859 to 1861. She later travelled to the USA and, at first, planned to study medicine there, but while Blackwell's degree was recognised, the loophole for foreign degrees to be accepted was now closed.

Jex-Blake was one of six women allowed to matriculate to Edinburgh to study medicine, but eventually they were not allowed to graduate and were instead offered certificates of proficiency which were not enough to allow them to be registered. She assisted in the establishment of the London School of Medicine for Women then when a hospital agreed to give clinical training in 1877, and after an act which she helped draft gave examining bodies the power to examine women, she used an MD from Berne and certification from Ireland, and was legally allowed to practise medicine in Britain.

Elizabeth Garrett Anderson became a nurse at the Middlesex Hospital and learned medicine on the job, attending lectures given for doctors, but she was sacked after complaints were laid. She had met Elizabeth Blackwell and wanted to emulate her, but another loophole had to be found first. Garrett found the Society of Apothecaries did not specifically bar females from taking their examinations, so she sat for, and passed the exams. Horrified, the apothecaries closed the loophole, but with financial help from her father, she established a medical practice in London, and in 1870, she obtained her MD from Paris.

Slowly, starting with Blackwell in 1859, the profession came to recognise that when you are fighting disease and death, all allies have to be welcome and women were allowed in, even where they were still not welcomed.

## DISEASE AND MORTALITY

Today, understanding evolution is second only to the germ theory in explaining how disease operates, so until the end of 1859, one of the cornerstones was missing. Over time, the operation of the principles Darwin outlined in late November had shaped not only the people of an urban world, but had also shaped the diseases they suffered from. Any strain of disease that killed too fast was likely to be replaced by other strains that acted more slowly, offering more chance to spread through the population. Going the other way, antibiotics which are used wrongly or foolishly have the potential to produce resistant bacteria, all following the evolutionary pattern Darwin described. On another front, there was selection for resistant humans as well.

Before germs came into the picture, epidemics rose and fell, often for no apparent reason. New Orleans had 4845 deaths from yellow fever in 1858, and just 91 in 1859. The mortality rate for the US Army soldiers who garrisoned the frontier forts in Texas between 1849 and 1859 was about 3.5 per cent each year, but living conditions were miserable and the soldiers suffered numerous bouts of disease (66,846 cases were reported among 20,393 soldiers). There were more than 20,000 cases of fevers, 14,000 cases of gastrointestinal diseases, and 7000 incidents of wounds and injuries.

London recorded 70 deaths in the third week of January including 26 infants, and 15 old people over 60. Half the deaths involved respiration: 11 cases of consumption, 12 bronchitis, seven pneumonia, four whooping cough, one asthma. Whooping cough and measles had been unusually fatal, especially in east London, but there had been no deaths from fever, smallpox, scarlatina and diarrhoea that week. It would not last.

Leprosy was on the rampage in Hawaii, where the steps taken show a clear understanding of infection. The Hawaiian government exiled those with leprosy to a coastal spot on Molokai, cut off from the rest of the island by the near impassable cliffs that reared 1800 feet behind the coast. Step by step, the pieces were falling into place, but there were still many ways to die, and precious few ways to save lives.

In Wales, a workman died after being bitten by a rabid cat. Cholera and rinderpest were in epidemic proportions in what is now Germany, cholera was loose in Costa Rica, and southern England had an outbreak of malaria, known there as ague. From 1840 to 1910, there were 8209 ague deaths in England, with Kent, Essex and Cambridgeshire the heaviest hit. In 1859 alone, there were around 370 deaths from malaria in Britain.

Nathaniel Hawthorne was dividing his time between Rome and Florence. His daughter Una caught malaria in October 1858, and was near death in April 1859, so the family returned to England while he stayed in Italy to finish *The Marble Faun*. The French and Austrian armies in northern Italy also suffered from the disease in 1859. In Vietnam, French troops were being laid low by malaria, a pattern that would be repeated: in the 1960s and 1970s, more Americans were stricken with malaria in Vietnam than were shot, though the number of American deaths from malaria was just 78, thanks to better drugs.

In the 1860s, there would be some 10,000 deaths during the US Civil War from malaria, but that was nothing new. Three generations earlier, one of the first expenditures by the Continental Congress of the US had been $300 for quinine for George Washington's troops—and British troops in North America were badly affected by malaria as well. As *Scientific American* commented in June:

> Fever and ague prevail every season, especially near the creeks and swamps of Long Island. One family claimed they avoided ague while others around them caught it, by having a good fire in the house on every 'chilly and damp night in summer and Fall'. Indians were reported to avoid chills and fevers when passing swamps at night or in the early morning by covering their nose and mouth with some part of their garment to warm the air.

In Costa Rica, cholera intruded into politics. Juan Rafael Mora Porras had been acclaimed as a hero after he ousted an American adventurer–invader called William Walker, but when a cholera epidemic killed 10 per cent of the population in 1859, chaos and social disruption led to economic problems, and Mora was blamed. He was forced out but tried to return, lost, and was shot by his former subjects on 30 September 1860.

John Snow might have died in 1858, but his influence lived on. James Reynolds had published the Snow map of London's Broad Street area in 1857, showing the position of the fatal pump, and the locations of the affected houses. The map is often presented as the tool Snow used to work out what was happening, but it was actually an after-the-event proof that his intuitive reaction was right.

Snow's map was drawn and engraved, hand coloured and reprinted with the date 1859, and it was then used to persuade doubters that cholera was caused by something in the water. It was a massive advance on 1830s theories that related cholera susceptibility to hair colour, or the ideas listed in Charles Kingsley's 1857 novel, *Two Years Ago*, where the smell of rotting fish and the hysteria induced by listening to non-conformist preachers were treated as possible causes, albeit with a passing reference to water being a possible carrier.

The world's 1857–59 influenza epidemic also hit the Pacific. The first foreign traders arrived in the Marshall Islands in 1859, bringing influenza with them. There were so many deaths that the survivors did not know what to do with the bodies. Missionaries said some bodies were just placed in the sea, wrapped in mats and fitted with small sails, in the hope they would blow away. The influenza problem continued, with measles and typhoid fever adding to the burden.

Smallpox was rife in Britain, though vaccination had been free to the poor since 1840, and compulsory since 1853. Many, but not all of those with the disease, had not been inoculated, but the existence of a few inoculated sufferers was the real worry. 'Two Paddington MDs' wrote to *The Times* in November, suggesting that it was time to renew the cowpox material used in inoculations, as it seemed to be losing its strength. Worse, taking cowpox from people in London carried the risk of passing on the 'poisons' of other diseases, such as syphilis and scrofula. Clearly, these two gentlemen had a good understanding of how at least some diseases might be transmitted, even if they did not accept germs as a reality.

In France, tobacco was blamed for cancer of the mouth in patients over 40, especially poor people who smoked short-stemmed pipes, unlike the rich, who chose long-stemmed pipes or cigars. Water pipes were used in Asia where this cancer was unheard of, making some believe the cancers were caused by heat, not nicotine. A German study showed that half the deaths of men between 18 and 28 years of age 'originate in the waste of the constitution induced by smoking'.

Some preventative treatments were truly fearful. A South Carolina newspaper told readers to clear a room of mosquitoes by taking a piece of gum camphor, a third the size of an egg and evaporating it over a candle or lamp, making sure it did not ignite. Even with the window open, the room remains clear of mosquitoes, readers were assured. Given that the toxic vapour is explosive and can be absorbed through the skin or by inhalation, the experiment is not recommended to the reader. Some of the other ideas being circulated were just as unfortunate, but ideas that were published were ideas that could be tested and judged. Science became something anybody could read about, consider, and play with.

## COMMUNICATING SCIENCE

The general enthusiasm for science and ideas about science offered an excellent way to sell books. By the 1830s, said Harriet Martineau, the middle classes were buying five times as many books on geology as they were novels. By the 1850s, geology was a profession and the amateurs' interest had moved on to living things, but the wonders and phenomena of modern chemistry and physics rated a mention from time to time.

The Sydney readers of *Crouch's epitome of news, and miscellaneous gleaner* could find curiosities inserted as fillers when a page lacked a few lines to complete it. In one issue, readers were told how gaslights might be lit by scuffing the feet on a carpet, or by standing on a chair with the legs in four glass tumblers (for insulation) while being rubbed with a muff, and then touching the metal gas jet while somebody else turned the gas on. Static electricity thus became a commonplace.

Larger and more ambitious works of natural history appeared, much of it coming from impeccable church-related sources. Charles Kingsley first became well known when he published a children's book on natural history, *Glaucus: or, The Wonders of the Shore* in 1855, while his friend P.H. Gosse published *Seaside Pleasures* in 1853, *The Aquarium* in 1854 and *Evenings at the Microscope* in 1859, following up with *The Romance of Natural History* in 1860. Gosse also published *Omphalos* in 1857 to express his opposition to the notion that species could ever change.

Margaret Gatty was a naturalist, the wife and daughter of Anglican clergymen, the mother of a large family (four sons and four daughters survived infancy) who found time to write morally uplifting stories about animals, birds and plants. Her 1859 opus, *Aunt Judy's Tales* by Mrs Alfred Gatty, is a typical piece of saccharine ghastliness and cloying morality, but this was the pap on which children were generally fed.

The Wardian case or terrarium became popular in the 1840s, and the aquarium in the 1850s. By the 1880s, Camille Saint-Saëns could include an aquarium in his *Carnival of the Animals*, because the aquarium was then mature technology, but the heyday of the aquarium pioneer was the 1850s, when not everybody understood what was needed, hence this filler in *Scientific American*:

> To keep your goldfish alive, feed them an occasional earthworm or a few angler's gentles, and add a bit of Anacharis to save having to change the water.

Far off in Australia, Samuel Hannaford wrote and published *Sea and river-side rambles in Victoria: being a handbook for those seeking recreation during the summer months*. This work is 120 pages, devoid of illustration other than dinkuses at the end of some chapters. It mentions some useful manuals, points to some interesting sites to visit near Melbourne and even includes a reference to Darwin ('Nat. Voy. 27'). Then it gets down to a few practicalities, based on the assumption that people will go to collect, not just to admire nature.

Hannaford offered a recipe for making artificial seawater, and a description of 'floating' marine algae onto paper that would only make sense to somebody who already knew how to do it. Hannaford explained that seaweeds may be sent through the mails in thin muslin, enclosed in either tinfoil or thin sheets of gutta percha. For collection, he suggested either a covered basket or a gutta percha bag. There is not a lot of practical or technical advice for the budding amateur naturalist.

The first issue of the 'new series' of *Scientific American* came out in mid-1859 and Webster's Dictionary records 1859 as the earliest date for the use of 'oceanography' (though the OED has it as 1883). It was a very good year for learned societies: the Chicago Academy of Natural Sciences was incorporated as the Chicago Academy of Sciences, after being established in 1856, the Entomological Society of Philadelphia was established, the British Ornithologists' Union was founded in 1858, and they began their quarterly journal, *Ibis*, in 1859. During the year, the Royal Society of Victoria was founded, and in Paris, Paul Broca started the Societé d'Anthropologie.

If natural history was safe in the hands of the divines and their wives, the men of stronger opinions bestrode the world of science and its communication. Dionysius Lardner, the probable parent of Dion Boucicault, was born in Dublin in 1793. His father wanted him to follow in the paternal footsteps and become a solicitor, but having graduated BA from Trinity College, Dublin in 1817 and after several unpleasant years of toil, he was appointed Professor of Natural Philosophy and Astronomy at University College, London. He held this position until 1840, when he became involved with a Mrs Mary Heaviside, and left with her for the USA, where a lecture tour gained him £40,000.

On Lardner's death in 1859, *Scientific American* wrote, 'In 1840 he came to the United States under a compulsory visit, with the young wife of a British captain, and the affair caused much comment at the time'. It is likely that just as his alleged son drew audiences to see him on the stage in Sydney because the audience had a taste for scandal, Lardner also benefited from being famous for being famous—or infamous.

Unlike today's tabloid and pulp-mag 'celebs', Lardner was good value. Between 1830 and 1844, he edited *Lardner's Cabinet Cyclopaedia,* a library of 134 volumes covering and popularising science. After 1845, he lived mainly in Paris, continuing to write about science. He was well enough regarded as an economist for Karl Marx to have mentioned him in *Das Kapital.*

Scientists and non-scientists were beginning to understand the world around them, starting to observe more closely. They kept on enquiring, and what they learned, they shared around the world. In newer, rawer societies, self-taught engineers were converting scientists' notions to practical inventions, but the notions had to come first. Up until the middle of the century, it depended quite a lot on the habits of some well educated rural English vicars and amateur enthusiasts.

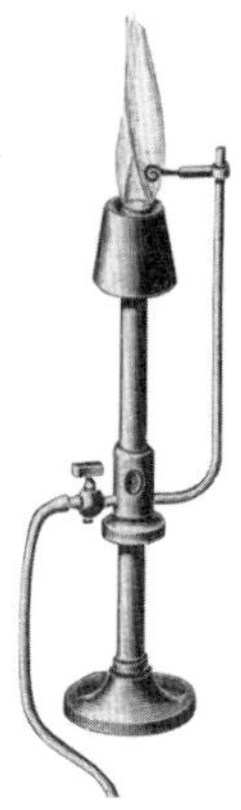

# 10:
# The rise of the professional scientist

*It freezes under people's beds.*

Gilbert White, Journal, 3 January 1768.

Gilbert White, vicar of Selborne and Jane Austen's near neighbour, had a gentleman amateur's fascination with weather records. So did John Dalton, schoolmaster and amateur chemist, who gave us the modern notion of the atom in 1808. Dalton kept a weather diary for 57 years until he died in 1844. Science in the 19th century was the play space of the gifted and curious amateur—and 'curious' sometimes took on more than one meaning.

William Buckland is a prime example. The son of a clergyman, he was ordained as a priest, but became an academic and practical geologist, the first Reader in Geology at Oxford, where he presented a close argument for the way geology demonstrated Biblical truths in 1820. Later, he was swayed by Agassiz' theories on Ice Ages and modified his stance, but he remained opposed to the idea of evolution, up to his death in 1856. Buckland was memorable, among other things, for eating all sorts of animals: zebra, snake, earwig, puppy, sea slug and even a bluebottle, though he declared mole the most disgusting thing he had ever consumed. He may or may not have eaten the dried heart of King Louis XIV (tradition says he did), but on his honeymoon, he identified some bones said to be those of St Rosalia as goat bones, and he investigated the alleged blood of a saint, which appeared fresh on a cathedral floor each morning. He lay on the floor, tasted it, and declared it to be bat urine (with which we assume he was familiar). They don't make scientists like that any more, but if he were alive today, he would surely be a leading television raconteur of science.

Gilbert White would today be an environmental blogger. He also corresponded with people like lawyer Daines Barrington who, among other things, examined the young Wolfgang Amadeus Mozart to see if the boy was a clever hoax, managed by the boy's father. Barrington suggested in a report to the Royal Society that the lad was not only genuine, but likely to be greater than Handel. Barrington also interviewed the last speaker of Cornish, Dolly Pentreath, and had some original ideas about fossils and polar exploration. White wrote to him about the keys in which owls hooted at Selborne.

Gilbert White died in 1793, Daines Barrington in 1800, at a time when the men of science were still expected to commit themselves to a range of endeavour and enquiry. By 1859, that era was over and the up-and-coming scientists had begun to specialise.

The names of those planning to attend the Leeds meeting of the British Association in 1858 were listed in *The Times*, ahead of the meeting. The list reads 'Sir David Brewster, Professor Faraday, Sir Roderick Murchison, Dr Whewell, Professor Wheatstone, Professor Airy, Sir William Hamilton, Sir Benjamin Brodie, Robert Stephenson, M. P., General Chesney, Mr. Hopkins, Mr. Darwin ...'

Darwin's name is followed by 33 others, few of them known today, even to those familiar with the period. Darwin was top-drawer, but not one of the truly great names like Astronomer Royal George Airy or William Whewell, the man who coined the word 'scientist' in 1840. Brodie (a surgeon who opposed amputating diseased joints), Chesney (Francis Rawdon Chesney, who surveyed a Suez Canal route in 1829) and Hopkins (probably William Hopkins, a mathematician, geologist and Cambridge coach), people barely heard of today, were all mentioned ahead of Darwin, but like him, many of them entered science through a back door.

Michael Faraday was a bookbinder's apprentice who read what he was binding, carried out some experiments, attended some lectures, and then joined the world of experimental science. Wheatstone first came to attention as a maker of musical instruments (and the inventor of the concertina) but he went on from there. Brewster trained as a clergyman, and in 1825, Darwin was a medical student at the University of Edinburgh but was so horrified by an operation performed on a child without anaesthetic that he gave up his studies without completing the course. Scientists were not trained before about 1850: they emerged.

In 1832, David Brewster, Michael Faraday and John Dalton, and botanist-explorer, Robert Brown (originally an army doctor), received honorary DCL (Doctor of Civil Laws) degrees at Oxford. Amateurs were recognised and celebrated, but they were not amateurs in the slighting sense we usually intend by the term today.

An amateur then was somebody who loved his subject, but by 1859, the role of the amateur in science had almost disappeared from view. The one exception was astronomy, where even today, amateurs are often the first to spot new comets.

## THE HEAVENS ABOVE

The Reverend Mr Thomas Webb pleased many amateur astronomers when he published the first edition of *Celestial Objects for Common Telescopes* in 1859. By 1917, it had reached its 6th edition,

The Transit Circle, Royal Observatory, Greenwich.

while my Dover edition was printed in 1962, more than a century after the first release.

Victorian amateurs loved their night skies. By 1859, Flat Earth notions were no more, and the heliocentric solar system with the Sun at the centre of a tiny corner of a much larger universe had replaced medieval notions that centred on us. In a filler piece in August, *Scientific American* explained that light from the Sun takes seven and a half minutes to reach the Earth.

By way of contrast, a cannonball would take 17 years to cover the same distance, said the writer, adding that the speed of light was about 200,000 miles a second, meaning light from the nearest fixed star took five years to reach us. Then came a curious comment reflecting an orthodoxy that was already being challenged: 'a remote visible star, created at the time of the creation of man may not yet have become visible to our system'.

A popular solution to the 6000 years problem was to assume a number of separate creations of life, with humans only appearing in the most recent round of creation. That seemed to account for older fossils and other embarrassing contradictions, but it was at best a poor work-around. When human remains were found with those of extinct animals previously assigned to earlier cycles, the whole scheme would fall apart. The skies were much on people's minds in 1859. There had been six comets in 1858, one of which rated a mention by Eliza Edwards, who apparently mistook the several instances for a single comet:

> that wonderful Comet—that there was So much Said about its coming in contact with, & destroying the earth—a year or so ago—has been visible here a week past—& tis by far the largest & most beautiful one Ive ever seen—I wish I knew whether you have seen it. It frightens the Kanaka's dreadfully—they cant Seem to understand about it.

*Scientific American* on the other hand, knew the comets were separate but still noted that '... those who had missed the comet were consoled that it would reappear in 2147 for a few nights'. There was more excitement for night watchers when Comet Tempel 1 appeared in 1859—and that particular body has been in the news again in recent years, after NASA's Deep Impact mission examined the same comet in 2005.

The ship *Southern Cross* left Boston on 10 June and arrived at San Francisco on 22 October, a passage of 134 days. They were 23 days off Cape Horn, and that was where passengers and crew saw an amazing auroral display on 2 September, thanks to a major solar storm. The storm was so violent that English astronomer, Richard Carrington, detected solar flares on the Sun, the first time they were seen.

Colourful auroras, usually only seen in polar regions, were visible at Rome and Hawaii. These were admired, but then the damage began. The storm sent a plasma blob hurtling out of the Sun, much faster than any cannonball, reaching the Earth in just under 18 hours. One day, another blob will come our way, but the damage next time will be far worse.

In 1859, telegraph wires suddenly shorted out across the US and Europe, causing fires in many places, but it was comparatively minor damage. A modern solar blast like that of 1859 will cost billions of dollars as phone lines, power lines and communications satellites and earth stations are fried. Even computers and home networks could be at risk.

A couple of weeks after the flares, *Scientific American* reported that the current generated in the telegraph wires was enough to overcome the telegraph batteries. In some cases these were shut off, after which ' ... messages were actually sent between Philadelphia and this city by the Aurora'. That is to say, the telegraph keys clattered aimlessly under the influence of stray currents induced in the lines.

Meteors were also popular with amateurs, who could collate reports that sometimes allowed a meteor's landing place to be identified. Nobody thought meteors or comets influenced the weather any more, yet it was an eminent professional astronomer who proposed that professional meteorology might work.

## WEATHER AND OTHER PHYSICAL EFFECTS

During the Crimean War, a storm hit Balaklava on 14 November 1854, when a British and French fleet was anchored in an exposed position. The storm drove the French *Henri IV* ashore and destroyed or damaged another 46 Allied ships. Both Britain and France were outraged that their combined Imperial might could be so challenged by the elements.

In 1846, Urbain Leverrier calculated the orbit of Neptune from the orbital quirks of other planets. He gave Johann Gottlieb Galle in Berlin his predicted coordinates to find the new planet. Galle found Neptune on his first try, just where the Frenchman said it would be, and Leverrier's name was made. People took notice when he said the Balaklava storm could have been predicted, 24 hours before it arrived, which would have allowed the ships to be moved to safety. That, at least, is the formal story.

Things are never quite that simple, and other investigators were seeking a science of weather even before that. Britain's

Meteorological Office was established in 1854, mainly to keep records. Robert FitzRoy, captain of the *Beagle* when Darwin sailed in her, served as Governor of New Zealand from 1843 to 1845, before he was recalled to England and promoted to admiral. In 1859, he was asked to set up a naval weather service.

Across the world, there was a proposal in the *Sydney Morning Herald* in 1859 for a storm warning system. Bad weather in Adelaide would trigger warnings in Melbourne and Sydney, as the storm rolled east. This would be mainly for the benefit of the 'nautical population' (and would not do much for Adelaide!).

The paper started providing basic meteorological data in April 1857, and this slowly increased. Sydney's official observations of the weather began at Observatory Hill in 1859, and in Vancouver, the Royal Engineers began recording weather data as well. The hobby of Gilbert White and John Dalton was finally a science.

The extremes of 1859 must have helped raise concerns about the weather. California suffered a severe heat wave, and for 10 days, the temperature at Sacramento rarely fell below 100 degrees. Paris had a heat wave in June, while July was the hottest in Paris since 1793, with the thermometer rising above 90 degrees, 12 days in a row. England reported a highest shade temperature for 60 years: 89 degrees. 'Of late years, the summer months of June, July, and August, appear to have had an increasing temperature', said *Scientific American*. This contrasted with January, when the *New York Times* reported that New York was 10 degrees below zero and Buffalo 20 below. The mercury rarely rose above zero, private carriages were hardly seen on the streets, and omnibus drivers took every opportunity to dismount, to avoid freezing to their seats.

Australia's climate was supposed to be sunshine and drought with occasional passing showers, said *The Times* in late March, but since the first hay crops were ready to cut, Victoria had been suffering severe weather. Hailstones from walnut size to six inches across were reported at Geelong in early February. They had smashed fences, stripped fruit trees and killed poultry, calves and even full-grown cattle.

Everybody talked about the weather, but what was needed was a theory, a science that could be used to make predictions. That was coming, but from an unexpected quarter. At the Vatican, a Jesuit priest was made director of the observatory at the Collegio Romano in 1849, after two years studying physics at Georgetown, Washington DC. Father Angelo Secchi is remembered today for suggesting the classification of stars by their spectra and for his Secchi disc, used to assess turbidity in water, but in 1859, he looked at air masses and how they moved. He found that 'atmospheric waves' (where a 'high' moves) could be detected, and that these waves could travel from Rome to London in about 36 hours.

At the American Association for the Advancement of Science meeting in August, Professor Joseph Henry spoke about weather theories. Some thought air flowed from the equator to the Poles, but he disagreed, citing Mr Wise the balloonist who said he had always drifted east in 200 ascents, once he reached the upper strata. Wise thought and hoped this effect might would carry him to Europe, though he conceded there might be a reversal in mid-ocean.

Henry believed that English meteorologists were hampered by getting no information from the west, while the east coast of the

US had telegraph lines extending west for thousands of miles, into the region where much of their weather originated.

In September, *Scientific American* explained how J. Kuechler of Gillespie county in Texas assessed the past climate. 'A tree bears its own witness,' he said. He felled three post oaks, two of them more than 130 years old, took a vertical section of each and planed it smooth. He varnished all three surfaces, and prepared tables showing the thicknesses of the rings which matched perfectly. He offered this as proof that rainfall alone causes the difference in annual rings. The record took him back as far as 1725, revealing 67 wet summers in 133 years, the rest being divided into dry, very dry and average seasons.

A recording thermometer was described in April, the work of a farmer called Gautlett from Middlesborough-on-Trent. He used a long tube made of thin sheet zinc, attached at one end only to a dowel of wood. As the temperature varied, the relative positions of the two ends also varied, allowing a pencil to move across a revolving cylinder containing a strip of paper, wound through by clockwork.

Weather studies had some curious influences on physics. John Tyndall's explanation of what we now call Tyndall scattering (why the sky is blue) arose from studies he began in 1859 and helped us understand light. The mercury barometer gave us the first captive vacuum, in the space above the mercury, and by 1859, a mercury air pump, designed by A. Gairaud was available for purchase. This was better than existing air pumps which could not generate a vacuum lower than 1⁄16 of an inch of mercury (around 0.2 per cent of atmospheric pressure). Now with just 20 or 25 pounds of mercury, all the usual experiments might be performed, though *Scientific American* suggested making part of the apparatus from that new wonder material, gutta percha. With the pump available, amazing things were about to happen.

In 1859, Johann Hittorf and Julius Plucker observed cathode rays in a vacuum tube and saw them bending in a magnetic field. They called their tube a gas discharge tube, but this was a little misleading: it needed a good vacuum, though it led on directly to neon lights, thermionic valves (vacuum tubes), light bulbs and television tubes.

Science was falling into the hands of professionals, people who not only specialised, but sought out each other's company.

## THE AGE OF PHYSICS AND CHEMISTRY

There is probably a doctoral dissertation or five in studies of the role of the tearoom in the advance of British chemistry and physics in the 19th and 20th centuries. New ideas required discussions with other professionals, and the laboratory tearoom provided a suitable place for discussions.

By 1859, most scientists worked in laboratories with others of a like mind, along with assistants. The gas discharge tube could only be developed with the help of pump makers, glass blowers, machinists and makers of fine insulated wire, so laboratories needed skilled artisans on hand (or in a few cases scientists who were themselves skilled artisans).

Chymistry was about to become chemistry, and natural philosophy was becoming physics. It would be 10 years before Dmitri Mendeleev worked out his periodic table, but by 1859, enough elements were known that people were beginning to discuss some of the patterns they could observe. Septimus Piesse found a pattern of sorts in the elements, and used it to massage and embroider his data in an article in *Scientific American*:

> From Oxygen being right, I believe the rest to be right; consequently, the atomic weight of Selenium to be 400, and not 494.5. If this be the case, then one of two things must necessarily exist—first, if Selenium is a true element, then its atomic weight is stated too high; or secondly, if it is an oxyd, it is a trifle too low, 6.42.

Piesse was a science writer rather than a chemist, an amateur out of his depth at a time when there was a premium on being professional. There *was* a pattern of sorts in the data, but it took a little more chemical sophistication to tease it out. Mendeleev went abroad to study in 1859 and heard Cannizzaro speaking of his work on atomic weights in 1860. This impressed him and set him thinking, leading to the periodic table in 1869, once he had enough data to work on.

The spectroscope was invented in 1859, by two German scientists, Robert Bunsen and Gustav Kirchhoff, who revealed that spectral lines could be used for analysis. By 1868, Jules Janssen would use spectra to detect helium in the Sun; Bunsen had predicted this application in November 1859, writing to Henry Roscoe:

> Thus a means has been found to determine the composition of the sun and fixed stars with the same accuracy as we determine sulfuric acid, chlorine, etc., with our chemical reagents. Substances on the earth can be determined by this method just as easily as on the sun, so that, for example, I have been able to detect lithium in 20 grams of sea water.

By 1864, William Huggins took the spectrum of a nebula, and before long, Doppler shifts were being measured on photographs of spectra, and we were on the way to the notions of expanding universes, Big Bangs and much more—all discoveries with sound roots in 1859.

Chemists had first dabbled and then worked systematically to get the first of the new aniline dyes that were coming onto the market in 1859, the year William Crookes founded the weekly *Chemical News*. He conducted it on much less formal lines than is usual with journals of scientific societies, editing it until 1906, training generations of professional chemists. Still, science was science in 1859, without any thought for some forms being greater than others. The cruel observation that there were two sorts of science, physics and stamp collecting, was still in the future.

## THE GENTLEMAN NATURALISTS

Any Web search on the string *<physics stamp collecting>* will reveal a wealth of variants, all attributed to Lord Rutherford. The variety suggests a certain apocryphal quality, but it is quite likely Rutherford did have and hold—and possibly even express—the opinion that there are two kinds of science: physics and stamp collecting, but that was in the 20th century (and just for the record, physicist Rutherford won a Nobel Prize—in chemistry!). In the 19th century, in all branches of science, gathering facts and then trying to make sense of them was a reasonably useful endeavour that, in many ways, began with an eighteenth century ship of that name.

H.M. *Bark Endeavour* was a Whitby collier, taken into the Royal Navy and sent off to Tahiti to observe a rare transit of Venus across the Sun in 1769. The journey was necessary because two sets of observations had to be taken from known distant places, allowing clever mathematics to be used to estimate the actual size of the solar system. As a bit of an afterthought, a rich amateur called Joseph Banks went along as a naturalist. This set a precedent for later scientific expeditions which would always include a supply of people skilled in collecting, identifying and preserving specimens.

Scientists sailed with French expeditions under La Pérouse and Baudin and with a number of British coastal expeditions around Australia. Others visited Africa or the Americas, and the voyage of the *Beagle* took Darwin around the world. The five-year US Exploring expedition surveyed the Pacific and the voyage of H.M.S. *Rattlesnake* carried TH Huxley to Australia.

Then there were the naturalists who went out on their own like Alexander von Humboldt, Alfred Russel Wallace, who goaded Darwin into publishing, and Wallace's friend, Henry Bates. For the most part, they were untrained or self trained, but their work shone because if they did not start out experienced and expert, they had the will and they certainly knew their craft by the time they finished. Thanks to the gentleman amateur naturalists, the museums of 1859 Europe were stuffed with specimens, and the professionals were still enriching the collections, filling them out.

The Museum of Natural History in Paris already had 36,000 plants on display, and the whole number of plant species was estimated to be at least four or five hundred thousand, ranging in size from a speck of mildew to the towering trees of Malabar, 15 metres in circumference, and banyans, covering two hectares. The raw material to tease out evolutionary relationships had been gathered and stored in Europe.

On top of that, some 60,000 animal species had been identified, including 44,000 insects, 4000 birds, 3000 fish, 3000 shellfish, 700 reptiles and 600 mammals. It was a store that made it easier for

Natural history exhibits were only some of the many offerings at museums.

the observer to see the patterns in the diversity of life, patterns that were the unmistakable trackways of past evolution.

There were also believed to be 80–100,000 species of tiny life forms, invisible to the naked eye. A general awareness of these was another key item in progressing toward a germ theory, and all of this evidence had been collected by genteel, but dedicated, amateurs.

Even royalty could play the scientific game. In 1859, the future emperor Maximilian of Mexico went collecting specimens, up the Amazon River. His brother, the Austrian emperor, had disapproved of Maximilian's liberal stance when he was regent in Milan, so Maximilian was relieved of his duties and went to the Americas. While he was there, the question of Maximilian becoming emperor of Mexico was first broached, but he preferred exploring.

He became emperor of Mexico in 1864 under a foolish scheme hatched by Napoleon III. When Maximilian was shot by Mexican forces in 1867, his embalmed body was taken back to Austria in SMS (Seiner Majestät Schiff) *Novara*, arriving in Vienna in early 1868. In the first few months of 1859, *Novara* was treated as a research vessel, even if attached to the Austrian navy.

Today, the idea of landlocked Austria having a navy may seem odd, but before World War I, the Austro-Hungarian empire had a significant Adriatic coastline. *Novara* was built at Venice, which was under Austrian control when the ship was started and again when she was completed in 1851, one of the last timber-hulled warships without iron plates. When her keel was laid in Venice in 1843, the Austrian name for her was *Minerva*, but Italian rebels captured the partial ship in 1848, and renamed her *Italia*.

This was a banned word under the Austrians, who later took Venice, and the ship, back and named her after an 1849 battle in which Austria had defeated Piedmont. She sailed for the Pacific as *Novara* in 1858.

When the ship returned to Trieste in August 1859, after leaving Australia in December 1858, Piedmont's ally Napoleon III ignored her provocative name and gave her the same neutral status French scientific craft had met in Australian waters in uncertain times at the start of the 19th century. Ferdinand von Hochstetter, the geologist who stayed in New Zealand to carry out mapping there, left his mark when he returned to Austria: the New Zealand Alps have a Mount Hochstetter and a Novara Peak, as well as a Dom Hochstetter. He arrived back in Trieste in 1860, and in 1861, married Georgiana Bengough, daughter of the English director of the Vienna city gasworks.

Some of the *Novara* science was fairly minimal. Schlumberger, a Viennese winemaker, provided Sekt (sparkling) wine to the *Novara* expedition, to show that their wines would travel. It was more a test platform than a research vessel. Among the innovations tried and publicised, the ship carried a Scottish dragnet for sampling the seafloor, a French water distillation plant, tinned food and enamel dishes. It was like modern sponsored expeditions, complete with product placement.

In a paid advertisement, the new aneroid barometer was praised by *Scientific American* as useful for explorers 'although we believe that the Smithsonian Institution does not approve of them'. It added '… they are small and compact—and cost only $10 when provided by Mr. Kendall of Massachusetts'.

Serious explorers had few comforts and many discomforts. Captain M'Clintock returned to London in 1859, having been sent out to look for any sign of the fate of Sir John Franklin and his men who had all perished in 1847 while searching for the fabled North–West Passage. He published an account of what he had seen and found, and late in 1859, presented a memoir of his findings to the Royal Geographical Society. One of his officers, Lt Cheyne, showed off some stereograms of some of their findings.

It was a good year to recall von Humboldt. 'His canvas was the universe, and he used his pencil with a master hand', wrote *The Times*. The last true polymath in the sciences, murmured the *Gentleman's Magazine*. He was the first European to see native South Americans preparing curare arrow poison from a vine, the first to warn of the need to preserve the cinchona plant, which was being drastically over harvested for the quinine in its bark, the first to describe vertical zonation in vegetation while moving up a mountain and the first to make accurate drawings of Inca ruins in South America.

Herschel Babbage was a gentleman scientist of great brilliance but limited political skill. He had surveyed railway routes, assisted his father Charles in building calculating machines and much more. Settling in South Australia, he went to explore the wilderness, and went fully equipped. His backers wanted him to find gold and agricultural land; he proceeded in a slow, meticulous and scrupulously scientific way. For this, he was relieved of his command by Peter Warburton, an inept but politically agile policeman. At a meeting of the Royal Geographical Society on 14 March, Babbage's work was first damned with faint praise:

> A large expedition had been started under Mr Babbage at a great expense to the colony, but it was hampered by its own weight, and did not attain any very considerable distance, although the country it passed over was thoroughly examined, both to the right and the left of the route, and carefully mapped by its leader.

Then came a detailed reading of Warburton's work, ending with his conclusions 'at the close of his numerous journeys', as he described it. His account betrays an abysmal ignorance of and lack of sympathy for the arid and fragile land he had passed through:

> Most of the country I have visited seems admirably adapted for pastoral occupation. It is one that would be greatly improved by being stocked, the surface would become firmer, and the thin coating of small stones would be just sufficient to prevent rapid evaporation, but not to interfere with the growth of grass, which would soon spring up under sheep. There is no scrub. I saw very few wild dogs, no kangaroos and no natives. Sheep might be run in flocks of several thousands, and I believe that for every single sheep the country could carry the first year, three might be put upon it in the third season. The ground is high, would be dry under foot when made firmer by the treading of the sheep, and it is clean for the wool. A little rain would leave plenty of temporary surface water.

Clearly, the ungifted amateur could do real harm, but the men of the Royal Geographical Society were swelled with pride. Said John Crawfurd, FRGS, of Warburton, Babbage and Stuart (who was about to set out again on his epic quest to find a route for the Overland Telegraph): 'They are true Englishmen, countrymen of the discoverers of the steam-engine, the locomotive, the electric telegraph ...'

England was also the country of the political backstab. Captain Sir Richard Burton was awarded the gold medal of the Geographical Society on his return from Tanganyika, 12 days behind John Speke, who, with a perpetual eye for the main chance, had sailed straight for England, where he arrived 9 May.

He at once took a very unfair advantage of Burton 'by calling at the Royal Geographical Society and endeavouring to inaugurate a new exploration' without his old chief. David Livingstone was in Malawi as part of the Zambezi Expedition, but Henry Morton Stanley, who would meet him and murmur 'Dr Livingstone, I presume', and go into the record books as the epitome of stiff-upper-lippedness, arrived in New Orleans as a cabin boy in 1859, before fighting on both sides in the US Civil War. Livingstone and Stanley staged their austere meeting for the benefit of the onlookers, and were actually a great deal more excited, but they held that it was proper for explorers to be stoic. It was ever thus.

In a piece on exotic meals, *Scientific American* wrote of the strange foods that are eaten in many parts of the world, and the travellers who tried them: 'Dr Shaw enjoyed lion; Mr Darwin had a passion for puma; Dr Brooke makes affidavit that melted bears' grease is the most refreshing potion'. Thomas Shaw was appointed in 1720 as chaplain to the English factory at Algiers on a salary of £100 per annum. He travelled widely in northern Africa and published a book of his travels in 1738.

The puma-eating Darwin is our friend Charles, who mentioned eating this animal in chapter VI of *Voyage of the Beagle*, where he quotes Shaw as saying that 'the flesh of the lion is in great esteem, having no small affinity with veal, both in colour, taste, and flavour'. This, said Darwin, was also true of the puma.

Sir Arthur de Capell Broke, sometimes called Brooke or Brook, was born in 1791. He spent a winter in Lapland and Sweden, and says in an 1827 book that bear flesh is a great delicacy in Lapland. An original member of the Travellers' Club, he later set up the Raleigh Club and became its president. In 1830 some members of the Raleigh Club, with Broke's approval, formed the Geographical Society which became the Royal Geographical Society in 1859.

There would be one more amateur exploring expedition in Australia, the lethal Burke and Wills expedition, then exploring would be taken away from the untrained, the amateur and the inexpert. Those would be left with little to play with other than natural history and fossil hunting.

But even entry into those fields now required some level of expertise; explorers needed enough theory to interpret and understand what they saw.

## GEOLOGY VERSUS THE 6000-YEAR EARTH

Hugh Miller, geologist and stern churchman, died in 1856. The next year, his widow made reference in a new edition of his *The Old Red Sandstone* to 'infidels' among the geologists, so clearly the lines were being drawn on the age of the Earth, well before Darwin published. A few geologists joined Mrs Miller in her doctrinaire and fundamentalist approach to the age of the Earth, but the professional geologists and the vast majority of trained scientists accepted that life had been on Earth far longer than the 6000 years that could be read into a literal reading of the Old Testament.

Collecting fossils had become a popular obsession among the middle classes, exposing more people to evidence of ancient life. Mary Anning was a self-trained professional by the time she died, still a young woman, from breast cancer in 1847. Living near Lyme Regis on the southern coast of England, she found and prepared fossils for rich collectors and museums. She is even commemorated in a verse that was once an assertion about the reality of fossils, but which most of us now know only as a tongue twister:

She sells seashells, by the seashore
And the shells that she sells, are seashells I'm sure.

Mary Anning probably learned some of her skills from her father who died when she was 10 or 11, but rich people flocked to see her and her collections, and to learn from her how she identified the material she found. Lady Harriet Silvester recorded that Anning had reached the peak of her profession at 25 'by reading and application'.

Fossils are curious things, and fossil experts are adept at detecting slight variations that reveal hidden secrets. Most fossils carry subtle clues in their shape, their form, where they lie, or what lies around them, but perceiving this only comes after looking at large numbers of fossils with a clever eye. That sort of insight does not necessarily help explain how a fossil came to be where it was, but it is a start. After that, you are left with a choice of logical reasoning, inference, supposition and wild fantasy. Many people, finding a conclusion they don't like, will denounce another's logical reasoning as wild fantasy, or hail a colleague's wild surmise as pure gold. Those who do this can sometimes be anti-scientists of the worst sort, but they can also be scientists.

The ire of mainstream geologists was roused by catastrophists —people who felt that all of geology might have been produced in several major catastrophes like Noah's flood. The normal geological view is called uniformitarian, meaning that conditions have been uniform over the eons, with geology caused by processes we can see today, with weathering, erosion, volcanoes and other ordinary events shaping the Earth.

Some of the modern opposition to asteroid theories that account for the 'end of the dinosaurs' stems from this same visceral reaction to any suggestion that catastrophes shaped the Earth. Rational geologists now tend to assume a sort of geological punctuated equilibrium, where normal conditions apply most of the time, with the occasional surprise. All the same, geological mavericks who stress tsunamis, asteroids and other catastrophes tend to be looked down on, even today, and this trend was very apparent by 1859.

Boucher de Perthes had been an extreme catastrophist, and he was an amateur at a time when geologists were professional. It took the mainstream scientists a while to trust him, but he just kept on, digging interesting human-made tools from deep chalk deposits in

France. In the end, the scientists came around to him and his finds, thanks mainly to Charles Lyell, who visited Boucher de Perthes' excavations in 1859 and came away convinced that the tools were not only real, but offered strong evidence that humans were older than supposed. Lyell presented a paper on the topic at the September meeting of the British Association and published *The Antiquity of Man* in 1863. After his visit, he wrote, 'That the human race goes back to the time of the mammoth and rhinoceros (Siberian) and not a few other extinct mammals is perfectly clear ...'

The dispute over flint tools in *The Times* in November turned on age. The tools lay far deeper than tombs which contained coins 2000 years old. The Biblical 6000 years was too short for the depths at which the flints occurred, unless you assume a change in conditions. A massive flood like Noah's might explain the deep burial, but as the geologists knew, chalkbeds are formed slowly by tiny organisms, not by floods.

'The discovery of these relics of a race which seems to have been of far greater antiquity than any that has been hitherto supposed to have inhabited our planet, involves many interesting and difficult questions,' wrote T.W. Flower.

A correspondent from Notting Hill mentioned London's new sewers in *The Times*, noting that a half-mile cut at Shepherd's Bush, 35 feet deep, part of the new London sewers, was to be extended all over London. The writer wondered if the Geological Society could appoint competent people to visit these cuttings and take note of the strata revealed.

The debate showed the opponents, the naysayers, as anti-Micawbers—Dickens' character spent his life hoping for something to turn up, they spent their lives scrutinising the works of science, hoping for something to turn down. 'Senex' was quick to propose a wild fantasy to account for the flints, and just like today's creationists, he based it on a garbled version of the works of the enemy. 'Mr Darwin tells us of cliffs in Patagonia full of fossils. Suppose such a cliff face fallen down over an Indian burying ground, and a river afterwards diverging from its bed depositing a variety of detritus, or drift, including minerals from the Andes,' he wrote in *The Times* on 5 December.

The debate was not straightforward, as physicists questioned the geological view. The planet was warm below ground, they said, and must be cooling, which meant it used to be hotter, because there was no apparent source of continuing warmth. If you went back far enough, the planet would have been too hot for life, and that set a limit to the time life had existed. The answer to this paradox was that internal radioactivity has kept the planet warm for billions of years, but in 1859, radioactivity was unknown.

British geologist Henry Clifton Sorby started ripple research in 1859. He used to say that people laughed at him for trying 'to examine mountains with microscopes', but his work was typical of the fine detail often required to tease out the truth. He examined the ripple patterns left in sedimentary rocks formed in shallow water. He assumed uniformity, that conditions when the rock was laid down were similar to those of today. From that base, he described how the rocks formed, and even reported the direction the waves and currents had travelled. It was detail at once far too fine and far too confronting for critics who felt their Biblical interpretations had just as much validity.

The flint tools argument might have been good training for the debate that would descend, like Noah's deluge, once Darwin's book came out—but his book was already out, released on 24 November. People just needed to read it first.

## 1859 AND THE DAWN OF EVOLUTION

The idea of evolution did not start with Charles Darwin. His grandfather Erasmus knew all about it, and tried to find an explanation of why it happened; so did Lamarck. Charles Darwin just came up with a logical model that would explain the observed facts—though even that is misleading. Alfred Russel Wallace reached similar conclusions, and in later editions of the book, Darwin identified a number of other people who had published 'his' idea before him, and not been noticed. But Charles Darwin set out a monster volume of evidence to support his notion of natural selection—and that made him special.

In a century and a half, his model of natural selection has been examined in many ways. It has always withstood the tests thrown at it: Darwin's explanation of the observed facts has never been shown to be false, but it has been extended by new knowledge. There is an anti-evolutionist myth that scientists have closed minds, and will not be swayed: they worship Darwin, and that is that. Any half-scientist knows that he or she would become world famous by being the first to disprove Darwin's explanation of evolution. No amount of hero worship would stop them publishing.

Evolution and our understanding of it has grown from many other directions. Darwin relied mainly on fossils and comparative anatomy. Today's evolution has absorbed a lot of extra information from genetics, and it is supported by independent evidence from biochemistry, DNA sequencing, biogeography and embryology.

Inheritance was obvious, but genetics, a systematic account which explained how inheritance was passed on, was known only to one man, Gregor Mendel. He was already carrying out experiments at Brno in 1859 and he would publish his results in the 1860s, but they would lie on library shelves, ignored, until the next century. All the same, plant and animal breeders made very Mendelian assumptions about inheritance, and what we now call genes, throughout the 19th century, but it remained a practical technology, rather than a formal science.

There is one other background item needed for us to understand the fuss over evolution: the 1859 standard idea of 'species', which was not quite the same as the one we use today. To some people, even some scientists, each species was fixed at Creation, and could never change, because it was not in its nature to do so. This matches human observation, since species do not change very much when viewed on our timescale. This is why Darwin felt compelled to address the idea of selection in domesticated species. It is also why, with no knowledge of genetics or mutations, Darwin needed to talk of 'sports' that arose in cultivated plants and animals. Back then, it was all he had to go on.

Mutations were a puzzle—even though Prince Leopold, the first haemophiliac in his family, turned six in 1859, and his bleeding tendency must have been obvious by then. Still, animal and plant breeders all knew how a species could be made to vary by selecting new mutant forms. James Boyd Senior of Chester, South Carolina, announced in December that he had found cotton of a 'new or peculiar species', what we would now call a mutant strain.

The bolls were very numerous, and it took just 30 of them to make a pound. Even today, it takes between 70 and 220 bolls to make a pound of cotton, with weights ranging between two and six grams each.

Boyd knew what to do. He had obtained a few seeds at first, and now had a considerable number, which he had achieved by keeping the plants separate from other cotton (which shows that he realised they would interbreed and so were, in our terms, the same species). He anticipated getting 1600 pounds of cotton per acre from the new strain. No *Scientific American* reader needed the selection process explained, and the use of 'species' was clear to them as well, though it differs from what we now understand.

In January 1859, *The Times* mentioned the exhibiting, at the Crystal Palace Poultry Show, of the Cochin China variety, noting that '... some of the buff and cinnamon-coloured cocks of this species are really of such immense size and powerful form as would almost justify a separate exhibition of each on its own merits'. Both Boyd's cotton and the exotic fowls were referred to as 'species' when we would now call them 'breeds' or 'strains'.

Linnaean taxonomy, the standard way of naming plants and animals, had been accepted for a century, and all 1859 biologists assumed that living things were related. The whole system of naming was based on levels of relationship, so members of the same genus were assumed to be more closely related than species in different genera of the same family. Evolution was implicit in the accepted naming system, and all that was missing was a good explanation of why evolution had happened.

## The same river twice

So to summarise: any opinion on evolution around 1859 involved several key issues: whether species were fixed or changeable; the question of whether or not extinctions happened; the age of the earth; and the number of times life had arisen. This is why all sorts of odd events, like finding human traces among the fossils of extinct mammals, or disproving spontaneous generation, were more important then than they would be now.

As Heraclitus said, we cannot bathe in the same river twice. Thanks to advances in science, we have lost track of most of those original threads, because they no longer matter. We accept that species change, that extinctions happen, that our planet is much older than humans, and that life originated just once.

Even before this was cleared up, many people accepted that life had radiated out from a smaller number of forms, and various writers had tried to explain how this might have happened. Darwin was just the first to state publicly a notion that a number of other people had already had: that over long periods of time, the pressure of natural selection could lead to a species becoming two species, each fitted to some special local condition or conditions.

Before Darwin, the charlatans who these days attack evolution with specious claims that their notions 'deserve equal time' had other targets. A Mr Prince, of Flushing, Long Island, wanted the American Association for the Advancement of Science in 1859 to recognise spiritualism and mentalism as a means of communication of knowledge. 'You may refuse spiritualism a hearing, but if spiritual science courts investigation and you evade it, the world will form its own opinion and my purpose will be answered,' he told them. He requested a committee of six to investigate the matter, but according to *Scientific American*, he was 'coughed down'. It was a more robust age.

Men of religion were by no means opposed to Darwin. The botanist Charles James Fox Bunbury visited Charles Kingsley in early December and noted Kingsley's opinion of *On the Origin of Species* in his diary: 'He talked much of Darwin's new book on species, expressing great admiration for it, but saying that it was so startling that he had not yet been able to make up his mind as to its soundness'. He also liked Darwin's use of domesticated animals as an example of the power of selection. Within a few days, though, he had made up his mind about its soundness and wrote to Darwin:

> From two common superstitions, at least, I shall be free, while judging of your book. 1]. I have long since, from watching the crossing of domesticated animals & plants, learnt to disbelieve the dogma of the permanence of species. 2]. I have gradually learnt to see that it is just as noble a conception of Deity, to believe that he created primal forms capable of self development into all forms needful pro tempore & pro loco, as to believe that He required a fresh act of intervention to supply the lacunas wh[ich] he himself had made. I question whether the former be not the loftier thought.

Soapy Sam Wilberforce, of course, attacked Darwin's idea with all his might. Disraeli later described him to Queen Victoria as 'a prelate … who is absolutely more odious in this country than Archbishop Laud'.

On the other hand, there was no blanket acceptance of Darwin's book by scientists, either. Louis Agassiz disagreed deeply, and astronomer Sir John Herschel called natural selection 'The Law of Higgledy-Piggledy'. As a physicist, the temperature of the Earth was still an impediment. Darwin welcomed Kingsley's support and added the following comment in later editions:

> A celebrated author and divine has written to me that 'he has gradually learnt to see that it is just as noble a conception of the Deity to believe that He created a few original forms capable of self-development into other and needful forms, as to believe that He required a fresh act of creation to supply the voids caused by the action of His laws.'

Darwin and Wallace disagreed about the evolution of the human mind, with Darwin believing natural selection could explain it. Wallace was a spiritualist, and wanted there to be some form of providence involved. Herbert Spencer wanted natural selection to not kill the fit, but force them to improve. He favoured a sort of Lamarckian kindness with no basis in science. What was lacking until the 1900s was a sound command of genetics that might explain how evolution operated, a command that encompassed mutations, population dynamics and a notion of dominant and recessive traits.

The needed science was there, at least in Mendel's mind, because the very design of his experiments shows he was setting out to prove, by detailed study, something which he already suspected with great conviction in 1859. In essence, Mendel already knew the sorts of results he would get, and his aim was to verify what he knew. If only he had shared those insights: then we could *truly* call 1859 the year the world changed!

## ENVOI

As the year closed, *On the Origin of Species* was reviewed favourably in *The Times* by an anonymous reviewer (we know now that it was T.H. Huxley). It was a good review, unsurprisingly. By 2 January, the initial print run of 1250 was long gone, and John Murray advertised in *The Times* that it was in its fifth thousand. On that day, *The Times* reported that the Bromley (Kent) volunteers were well supported, with almost £1000 raised. One of the £50 donors was Mr C.R. Darwin.

It was a decent sum from somebody who now realised that he was about to be well off. A month earlier, *The Times* carried this advertisement which puts Darwin's £50 in perspective:

> Wanted for the Wickwar Grammar School [girls] a Certificated Mistress, who will be permitted to receive two boarders. Salary £35 per annum, with a residence. Copies of testimonials must be sent to the Mayor of Wickwar, on or before the 24th December,

> 1859. Copies of the rules will be supplied by the Mayor in answer to pre-paid application.

The year 1859 had brought some disappointments, said *The Times* at year's end. The *Great Eastern* had not reached America and no balloon had crossed the Atlantic, though an amazing trip had been made by La Mountain and Haddock. The Atlantic cable 'has become mythical', and it looked as though it would be some years at best before anybody tried to lay one again. But the use of shorter runs, either across to Siberia or through Iceland to Europe, seemed more likely.

The year had seen the deaths of Brunel, Stephenson, von Humboldt, Lardner and Nichol, as well as Prescott and Irving, none of those surnames requiring the qualification of a given name in 1859.

War was in the air, and the Council of the Society of Arts had put off repeating the Exhibition of 1851 until 1862, rather than 1861. France and Austria had been fighting in Lombardy, Europe was all under arms, the Kingdom of Two Sicilies was apparently about to rebel, Spain was, for the first time in centuries, at war (in Morocco) and England was re-arming for self defence. If the Indian Mutiny was at an end, said *The Times*, there was likely to be war in China and 'by the last intelligence, General Harney was doing his best at St. Juan to embroil us with our friends of the United States'.

It was the best of times, it was the worst of times, but after 1859, the times would never be the same again.

# References

*My quarrel with him is, that his works contain nothing worth quoting; and a book that furnishes no quotations, is me judice, no book—it is a plaything.*

The Reverend Doctor Folliott in Thomas Love Peacock's *Crotchet Castle*, I. 1831–37.

I don't know how many quotations I may have furnished for others, but I have certainly drawn on enough works that have offered quotations and illustrations to me.

I also referred to and dabbled in quite a few books and websites. The following are some of the highlights, sources I would recommend for further reading.

## PRINT SOURCES

Anon., *Report from the Select Committee on the Condition of the Working Classes of the Metropolis*. Sydney: Thomas Richards, Government Printer, 1859–60.

Anon., *The Victorian cricketers' guide*. Melbourne: Sands, Kenny and Co., 1859–1862.

Basalla, George, William Coleman and Robert H. Kargon (eds.), *Victorian Science*. New York: Anchor Books, 1970.

# REFERENCES


Baudelaire, Charles, 'The Salon of 1859', in *Art in Paris 1845–1862: reviews of salons and other exhibitions*, translated by Jonathan Mayne, London: Phaidon Press, 1965.

Baynes, E. Neil, *An account of the wreck of the 'Royal Charter', 26th October, 1859*. London: Printed for the author by Wyman & Sons, [1936].

Bektas, Yakup, 'The Sultan's Messenger: Cultural Constructions of Ottoman Telegraphy, 1847–1880', *Technology and Culture* 41(4), October 2000, pp. 669–696.

Best, Geoffrey, *War and Society in Revolutionary Europe*, 1770–1870. Thrupp: Sutton Publishing, 1998.

Boswell, James, *Life of Johnson, Volume 5, Tour to the Hebrides (1773) and Journey into North Wales (1774)*. Project Gutenberg EText-No. 10451.

Budd, W., On intestinal fever. *Lancet* 1859; ii: 4–5, 28–30, 55–6, 80–2.

Burkhardt, Frederick (ed.), *Charles Darwin's letters: a selection, 1825–1859*. Cambridge: University of Cambridge Press, 1996.

Cole, G.D. and Wilson AW, *British Working Class Movements, Select Documents*. London: Macmillan, 1951.

Crawfurd, John, *Proceedings of the Royal Geographical Society*, Eighth meeting, 11 March, 1859.

Crouch, G. J., *Crouch's epitome of news, and miscellaneous gleaner*. Sydney: G.J. Crouch.

Darwin, Charles, *On the Origin of Species*, first edition. London: John Murray, 1859.

Darwin, Charles, *On the Origin of Species*, sixth edition. London: John Murray, 1872.

Defoe, Daniel, *A Journal of the Plague Year*. Project Gutenberg EText-No. 9863.

Dracobly, Alex, 'Ethics and Experimentation on Human Subjects in Mid-Nineteenth-Century France: The Story of the 1859 Syphilis Experiments', *Bulletin of the History of Medicine* 77(2), Summer 2003, pp. 332–366.

Ellis, Roger, *Who's Who in Victorian Britain*. London: Shepheard-Walwyn, 1997.

Fowler, Frank, *The wreck of the 'Royal Charter', compiled from authentic sources with some original matter by Frank Fowler.* London: Sampson Low, Son and Co., 1859.

Fussell, G.E., James Ward R.A.: *Animal Painter, 1769–1859, and His England.* London: Joseph, 1974.

Garfield, Simon, *Mauve.* London: Faber and Faber, 2000.

Gates, Barbara T., *Victorian Suicide: Mad Crimes and Sad Histories.* Princeton: Princeton University Press, 1988.

Grayson, Donald K., 'Nineteenth-century explanations of Pleistocene extinctions: A review and analysis', in *Quaternary Extinctions: A Prehistoric Revolution*, ed. Paul S. Martin and Richard G. Klein. Tucson: The University of Arizona Press, 1984.

Groom, Barry, *Sydney in the 1850s.* Sydney: Macleay Museum, University of Sydney, 1982.

Hannaford, Samuel, *Sea and river-side rambles in Victoria: being a handbook for those seeking recreation during the summer months.* Geelong: Heath & Cordell, 1860.

Harvey, Robert, *Cochrane, the life and exploits of a fighting captain.* London: Constable and Robinson, 2002.

Hobbes, Thomas, *Leviathan.* Project Gutenberg file lvthn10.txt.

Hulbert, Archer B., *The Paths of Inland Commerce.* Project Gutenberg file tpoic10.txt.

Innes, Eliza Clunes, *A snapshot of Lake Innes 1859: the diary of Eliza Clunes Innes* (transcription by Arnold Mitchell). Duckenfield, N.S.W.: Arnold Mitchell, 2001.

Johnson, Miss, *Geography With Useful Facts for the Junior Classes in Schools.* Sydney: Sands, 1859.

Johnson, Samuel, *A Journey to the Western Isles of Scotland.* Project Gutenberg EText-No. 2064.

Keay, John, *Mad About the Mekong.* London: Harper, 2005.

Kennedy, A., *An authentic account of the wreck of the 'Royal Charter' steam clipper on her passage from Australia to Liverpool, October 26th, 1859: with an interesting addition of subsequent events and incidents, written during a residence at Moelfra, the scene of the catastrophe*. Dublin: M'Glashan & Gill, 1860.

Kirsch, George B., 'American Cricket: Players and Clubs Before the Civil War', *Journal of Sport History*, 11(1) (Spring, 1984), http://www.aafla.org/SportsLibrary/JSH/JSH1984/ /jsh1101c.pdf.

Lyell, Sir Charles, *The Antiquity of Man*, first published 1863. London: J. M. Dent, 1914.

Macinnis, Peter, *Australia's Pioneers, Fools and Heroes*. Sydney: Pier 9, 2007.

Mason, Peter, *Cauchu: the Weeping Wood*. Sydney, Australian Broadcasting Commission, 1979.

Masters, W. E., *The Semaphore Telegraph System of Van Diemen's Land*. Hobart: Cat and Fiddle Press, 1973.

McLynn, Frank, *1759: the year Britain became master of the world*. New York: Grove Press, 2004.

Miller, Hugh, *The Old Red Sandstone*, seventh edition, 1857. London: J. M. Dent, n.d.

Moorhead R., 'William Budd and typhoid fever'. *Journal of the Royal Society of Medicine*, 2002, 95: 561–4

Mossman, Samuel, *Narrative of the shipwreck of the 'Admella', inter-colonial steamer, on the southern coast of Australia: drawn up, from authentic statements furnished by the rescuers and survivors*. Melbourne: Printed and published for the Committee of the 'Admella' Fund by J.H. Moulines and Co., 1859.

Osborne, Milton, *Mekong* (revised edition). Sydney: Allen and Unwin, 2006.

Parkin, Andrew, 'Boucicault, Dion' in *Victorian Britain, An Encyclopedia*, ed. Sally Mitchell. London: Garland, 1987.

Parkinson, C. Northcote, *British Intervention in Malaya*, Singapore: University of Malaya Press, 1960.

Pool, Daniel, *What Jane Austen Ate and Charles Dickens Knew*. London: Robinson, 1998.

Reynolds, Maurice, *Railways; and how they were managed fourteen years ago: being the substance of a lecture delivered at the Temperance Hall, on Thursday, 20th October, 1859*. Sydney: John Ferguson, Publisher, 1872.

Ritchie, J. Ewing, *London*. Published 1860, http://www.victorianlondon.org/publications/aboutlondon-17.htm.

Rockmore, Daniel N., *Stalking the Riemann hypothesis: the quest to find the hidden law of prime numbers*. New York: Pantheon Books, c2005.

Schiffer, Michael B., 'The Electric Lighthouse in the Nineteenth Century: Aid to Navigation and Political Technology', *Technology and Culture* 46, Number 2, April 2005, pp. 75–305

Scholes, Percy A., *The Oxford Companion to Music*, 9th edition. Oxford: Oxford University Press, 1955

Schwartzberg, Beverly, 'Lots of them did that': desertion, bigamy, and marital fluidity in late-nineteenth-century America. *Journal of Social History*, 22 March 2004

Shridharani, Krishnalal, *The Story of Indian Telegraphs: a century of progress*. New Delhi: Government of India Press, 1953.

Standage, Tom, *The Victorian Internet*. New York: Walker and Company, 1998.

Takeshi, Toyoda, *A History of Pre-Meiji Commerce in Japan*. publisher unclear, 1969.

Taylor, Peter, *An End to Silence*. Sydney: Methuen, 1980.

Thomas, Edward, *George Borrow The Man and His Books*. Project Gutenberg EText-No. 18588.

Thomson, David, *Europe Since Napoleon*. Harmondsworth: Pelican Books, 1976.

Webb, Beatrice and Sidney, *The History of Trade Unionism*. New York, A.M. Kelley, 1965.

Wilkes, G. A., *The Stockyard and the Croquet Lawn*. Melbourne: Edward Arnold, 1981.

Winchester, Simon, *The Surgeon of Crowthorne*. Melbourne: Penguin Books, 1999.

## WEB SOURCES

Where I was aware of a Project Gutenberg source for a work, I have cited that under 'Print sources', on the principle that readers will be looking for books in the first instance. There are probably many more of the works I have used which are also to be found there. See **http://www.gutenberg.org/catalog/** as a good starting point.

Blinderman, Charles, The Huxley files, **http://aleph0.clarku.edu/huxley/**

Currency comparisons were based on data at **http://measuringworth.com/**

Darwin Correspondence Project, **http://www.lib.cam.ac.uk/Departments/Darwin/index.html**

To cross over from sources to source acknowledgements, I am eternally grateful to those who went before, in particular the mainly unnamed journalists at *The Times* and *The Scientific American* who reported what they saw. I accessed *The Times* online database, courtesy of the State Library of NSW and *The Scientific American* through Cornell University's 'The Making of America' collection.

I have seen the future and, if we are to have resources like this, it is good.

## PICTURE CREDITS

Corbis: p.3, p.8, p.48, p.73, p.98, p.109, p.137, p.146, p.161, p.242

Photolibrary / Bridgeman: p.4, p.14, p.168, p.180, p.190, p.196, p.221, p.231, p.260, p.276, p.287

*The Scientific American*: p.23, p.35, p.54, p.64, p.82, p.112, p.117, p.129, p.131, p.212

*The Lady's Book*, Louis A. Godey: p.184

*Illustrated London News*: p.255

# Acknowledgements

Many friends assisted me in the making of this book. At the head of the list, I must place Sue Watkins in California, a tireless seeker-out of unusual press clippings who pointed me in many directions at various times (and sometimes at the same time). Had I not been conscious of the publisher's expectations, I could have produced a book three times as long. Her eye for the obscure but interesting, the germane and the crucial kept me fed with material.

My favourite libraries, the State Library of NSW and the University of Sydney's Fisher Library, provided me with a great deal of curious material, especially the State Library's Mitchell Library, where some of David Scott Mitchell's eclectic collections of pamphlets afforded me a number of serendipitous delights. A book like this is necessarily the harvest of a quiet eye, but it depends on those who prepared the field in the first place.

Members of two lists, Stumpers-Talk and Upper Branches, fed me a range of interesting leads. Gloria Burns, Dee Churchill, Mike Pingleton, Glenda Pearson, Timothy Pwee, Mary Pritchard, Mary Lou White, Sylvia Milne, Charles Early and Roberta Gerber all helped with ideas and pointers.

Graeme Rymill in the library of the University of WA, known to me through Project Wombat, came up trumps for me in identifying the leophagous Shaw and the ursiphagous Brooke. He always does! Chris Lawson also knew something of Shaw, and shared. John Germain of PW found some lowdown on travel speeds in the eighteenth century. Jenny Michelle Riker, Tim Unwin, Pat Dodson and John Heaviside were people who, when approached by email, were happy to share snippets that helped me track down answers to my occasionally bizarre questions.

My wife Christine became something of an expert on 1859 as well, as evidenced by the way I could ask her if she would care to hazard a guess at the date of some odd event, and she could unerringly answer '1859'. She remained composed and supportive through all of that.

Among my readers, David Allen of Yeppoon, Queensland, must take the prize for scrupulous checking and querying of logical fallacies. I have to note that I revised a deal after David had finished; if any glitches remain present, I feel confident that I must have inserted them at a late stage. Other helpful readers whose efforts I may have similarly circumvented include my son Angus, Anne Hansen, Chris Forbes and Luke Owens.

My thanks to all of the friendly and supporting people at Pier 9, especially Diana Hill, but even more especially, the amazingly patient and painstaking Shelley Kenigsberg, who kept me on the straight and narrow and stopped me assuming that the reader could also read my mind.

It is traditional for authors to note that in spite of all the checking by others, the remaining blemishes are their fault, and their responsibility. I fondly hope there *are* no remaining blemishes. There may be what the careful reader considers to be idiosyncrasies, and that would not surprise me. I cheerfully acknowledge those as mine alone.

Over the years, I have written many books, but none has given me such fun, and none has brought me so close to such a rich cast of eccentrics. I plead that perhaps a little of it has rubbed off as I pursued my characters down their obscure burrows and hauled them back into the light.

My final thanks, then, to the people who lived and breathed 1859 and reality, and gave me an interesting place to live and breathe (virtually) for 21 months and three days from the night when it struck me that 1859 might be a suitable case for treatment.

# Index

# INDEX

First published in 2008 by Pier 9, an imprint of Murdoch Books Pty Limited

Murdoch Books Australia
Pier 8/9
23 Hickson Road
Millers Point NSW 2000
Phone: +61 (0) 2 8220 2000
Fax: +61 (0) 2 8220 2558
www.murdochbooks.com.au

Murdoch Books UK Limited
Erico House, 6th Floor
93–99 Upper Richmond Road
Putney, London SW15 2TG
Phone: +44 (0) 20 8785 5995
Fax: +44 (0) 20 8785 5985
www.murdochbooks.co.uk

Chief Executive: Juliet Rogers
Publishing Director: Kay Scarlett

Commissioning Editor: Diana Hill
Project Manager: Emma Hutchinson
Editor: Shelley Kenigsberg
Design Concept: Katy Wall
Design Layout: Susanne Geppert
Production: Kita George



National Library of Australia Cataloguing-in-Publication Data
Author: Macinnis, P. (Peter)
Title: Mr Darwin's incredible shrinking world / Peter Macinnis.
ISBN: 978 1 7419 6279 6 (pbk.)
Notes: Includes index.
Bibliography.
Subjects: Science– History.
Discoveries in science.
Technology and civilization.
Dewey Number: 509

A catalogue record for this book is available from the British Library.

Printed by i-Book Printing Ltd in 2008. Printed in China.